RaumFragen: Stadt – Region – Landschaft

Reihe herausgegeben von
O. Kühne, Saarbrücken, Deutschland
S. Kinder, Tübingen, Deutschland
O. Schnur, Berlin, Deutschland

Im Zuge des „spatial turns" der Sozial- und Geisteswissenschaften hat sich die Zahl der wissenschaftlichen Forschungen in diesem Bereich deutlich erhöht. Mit der Reihe „RaumFragen: Stadt – Region – Landschaft" wird Wissenschaftlerinnen und Wissenschaftlern ein Forum angeboten, innovative Ansätze der Anthropogeographie und sozialwissenschaftlichen Raumforschung zu präsentieren. Die Reihe orientiert sich an grundsätzlichen Fragen des gesellschaftlichen Raumverständnisses. Dabei ist es das Ziel, unterschiedliche Theorieansätze der anthropogeographischen und sozialwissenschaftlichen Stadt- und Regionalforschung zu integrieren. Räumliche Bezüge sollen dabei insbesondere auf mikro- und mesoskaliger Ebene liegen. Die Reihe umfasst theoretische sowie theoriegeleitete empirische Arbeiten. Dazu gehören Monographien und Sammelbände, aber auch Einführungen in Teilaspekte der stadt- und regionalbezogenen geographischen und sozialwissenschaftlichen Forschung. Ergänzend werden auch Tagungsbände und Qualifikationsarbeiten (Dissertationen, Habilitationsschriften) publiziert.

Reihe herausgegeben von

Prof. Dr. Dr. Olaf Kühne
Universität Tübingen

PD Dr. Olaf Schnur
Berlin

Prof. Dr. Sebastian Kinder
Universität Tübingen

Weitere Bände in der Reihe http://www.springer.com/series/10584

Vera Köpsel

New Spaces for Climate Change

The Societal Construction of Landscapes
in Times of a Changing Climate

Springer VS

Vera Köpsel
Institute of Marine Ecosystem
and Fishery Science
University of Hamburg
Hamburg, Germany

Dissertation at the Faculty of Mathematics, Informatics and Natural Sciences, Department of Earth Sciences, University of Hamburg, 2017

RaumFragen: Stadt – Region – Landschaft
ISBN 978-3-658-23312-9   ISBN 978-3-658-23313-6   (eBook)
https://doi.org/10.1007/978-3-658-23313-6

Library of Congress Control Number: 2018953000

Springer VS
© Springer Fachmedien Wiesbaden GmbH, part of Springer Nature 2019

This Springer VS imprint is published by the registered company Springer Fachmedien Wiesbaden GmbH part of Springer Nature
The registered company address is: Abraham-Lincoln-Str. 46, 65189 Wiesbaden, Germany

Für Opa und Oma

Contents

Abstract

This study investigates the role of different societal constructions of landscapes in local processes of climate change adaptation. Coastal landscapes are changing due to the impacts of climate change: increased wave action, extreme weather events, and erosion change the physical-material fabric of coastal places. Research on the societal dynamics of adaptation to climate change has advanced during recent years from technical and economic foci to considering individual perspectives and personal values. In this context, a growing literature on the relationship between place attachment and climate adaptation has emerged. Nevertheless, critical limitations are evident in the conceptualisation of people–place relationships within this literature; in particular, individual place constructions and their implications for processes of climate adaptation are given insufficient attention. Alongside the idea of place, a concept prominent in human geography for approaching people's perceptions of and relationships with particular physical spaces is that of landscape. In order to better understand the locally specific societal values that shape processes of climate change adaptation, I mobilise social-constructivist landscape research to investigate empirically how local actors in landscape management perceive the physical spaces in which they live and which they manage, and how these perceptions influence their perspectives on adaptation.

This study's research area is Cornwall, a rural peninsula at the south-westerly tip of England. Within Cornwall, particular focus lies on Godrevy headland and local processes of adaptation to coastal erosion. Renowned for its scenic landscapes and already affected by impacts of climate change, Cornwall is a county where the issue of landscape plays an important role in the area's economic activities and regional identity. Applying a qualitative methodology of semi-structured and so-called walking interviews, I uncover five contrasting narratives of Cornwall's landscapes among local actors in landscape management: an official policy narrative, along with perspective on the landscape as human-environment interaction, natural systems, visual beauty, and functional spaces of (agricultural) production. Analysed at the societal, sub-societal and individual level, each of these narratives results in a distinct perspective on climate change adaptation.

The implications of the landscape narratives become particularly evident in the case of Godrevy, where local actors negotiate physical-material landscape change in response to proceeding coastal erosion. To better understand the place-based values that shape their decision-making, I bring together human geographical debates about the human-nature relationship, landscapes as common goods, and the question of physical access to them; topics that decisively shape local discussions about adaptive landscape management at Godrevy. In addition, a focus on the individual accounts of local decision-makers sheds light on the interplay between professional knowledge and personal experiences in processes of climate adaptation.

In summary, this thesis makes important theoretical, methodical and empirical contributions to human geographic, values-based climate adaptation research; setting particular focus on societal constructions of the physical spaces in which climate change takes place. Especially the case of Godrevy thereby constitutes a valuable in-depth study that illustrates the substantial value of walking interviews for researching societal processes of climate adaptation in their local contexts on the one hand, and for understanding the place-based values that shape local actors' decision-making about physical-material landscape change in response to the changing climate on the other.

Kurzzusammenfassung

Die hier vorgestellte Studie untersucht empirisch die Rolle unterschiedlicher ge-
sellschaftlicher Konstruktionen von Landschaften in lokalen Prozessen der
Klimaanpassung. Viele Küstenlandschaften wandeln sich durch die Auswirkungen des Klimawandels: erhöhte Wellenaktivität, Extremwetterereignisse und
Erosion verändern die physisch-materielle Struktur der Küste. Wissenschaftliche
Forschung über die gesellschaftlichen Dynamiken der Anpassung an solche Kli-
mafolgen hat sich in den vergangenen Jahren von technischen und ökonomischen
Analysen dahingehend fortentwickelt, die individuellen Perspektiven und persönlichen Werte zu betrachten, die Anpassungsprozesse formen. In diesem Zu-
sammen-hang ist ein Fundus an Literatur entstanden, der sich mit den Zusam-
menhängen zwischen Ortsbindung und Klimaanpassung befasst. Diese Literatur
weist allerdings Schwächen auf, was die Konzeptionalisierung von Mensch-Ort-
Beziehungen anbelangt; insbesondere individuelle Wahrnehmungen von Orten
und deren Einflüsse auf lokale Prozesse der Klimaanpassung erfahren zu wenig
Berücksichtigung. Neben dem Konzept von Orten bzw. ‚Places', ist das Konzept
von Landschaft prominent in der Humangeographie, um die Wahrnehmungen
der Menschen von und ihre Beziehungen mit bestimmten physischen Räumen zu
erfassen. Um das bestehende wissenschaftliche Verständnis der ortsbezogenen
gesellschaftlichen Werte zu verbessern, die Klimaanpassungsprozesse beeinflussen, ziehe ich sozial-konstruktivistische Ansätze der Landschaftsforschung heran
und untersuche empirisch, wie lokale Akteure im Landschaftsmanagement die
physischen Räume wahrnehmen, in denen sie leben und die sie managen, und
wie diese Wahrnehmungen in unterschiedliche Perspektiven auf Klimaanpassung resultieren.

   Das Untersuchungsgebiet dieser Arbeit ist Cornwall, eine ländliche Halbin-
sel an der Spitze Südwest-Englands. Innerhalb Cornwalls liegt ein besonderer
Fokus auf der Landzunge von Godrevy und den dortigen Anpassungsprozessen
an Küstenerosion. Bekannt für seine malerischen Landschaften und bereits heute
von Auswirkungen des Klimawandels betroffen, ist Cornwall eine Region in der
das Thema Landschaft sowohl für ihre ökonomischen Aktivitäten als auch für
die regionale Identität eine bedeutende Rolle spielt. Unter Verwendung eines
qualitativen methodischen Ansatzes mit semi-strukturierten und so-genannten
‚walking' Interviews identifiziere ich fünf konträre Narrative über Cornwall's
Landschaften: ein offizielles Policy-Narrativ sowie Perspektiven auf die Landschaft als Mensch-Umwelt-Interaktion, Natursystem, visuelle Schönheit, und

funktionalen Raum für (agrarische) Produktion. Auf gesellschaftlicher, teil-ge-
sellschaftlicher und individueller Ebene analysiert, resultiert jedes dieser Narrative in konkrete Sichtweisen auf Klimaanpassung.

Die Implikationen dieser Narrative werden besonders deutlich am Beispiel von
Godrevy, wo lokale Akteure zurzeit eine physisch-materielle Neugestaltung der
Landschaft als Antwort auf Küstenerosion verhandeln. Um besser zu verstehen,
welche ortsbezogenen Werte ihre Entscheidungen beeinflussen, verbinde ich hu-
mangeographische Debatten um Mensch-Natur-Beziehungen, Landschaften als
Gemeinschaftsgüter, und die Frage nach physischem Zugang zu diesen Landschaften; Themen, die die lokalen Diskussionen um adaptives Landschaftsmanagement in Godrevy entscheidend prägen. Darüber hinaus beleuchtet ein Fokus
auf die individuellen Perspektiven auf Landschaft unter den lokalen Akteuren
das Zusammenspiel von professionellem Wissen und persönlichen Erfahrungen
in der Entscheidungsfindung um Anpassungsmaßnahmen.

Insgesamt leistet diese Arbeit wichtige theoretische, methodische und empirische Beiträge zum Feld der wertebasierten Klimaanpassungs-Forschung und
setzt dabei einen besonderen Schwerpunkt auf die gesellschaftlichen Konstruktionen der Orte, die vom Klimawandel verändert werden. Insbesondere das Beispiel Godrevy ist eine tiefgreifende Fallstudie, die einerseits den Mehrwert von
‚walking' Interviews bei der Erforschung gesellschaftlicher Prozesse von Klimaanpassung verdeutlicht und andererseits zeigt, welche ortsbezogenen Werte die
Entscheidungen lokaler Akteure in Bezug auf die physisch-materielle Anpassung
von Küstenlandschaften an den Klimawandel prägen.

List of Abbreviations

AONB	Area of Outstanding Natural Beauty
CC/cc	Climate Change
CCFF	Cornwall Community Flood Forum
CoaST	Cornwall Sustainable Tourism Network
Coun	Cornwall Council (in interview analysis)
Counc'lor	Cornwall Councillor (in interview analysis)
CSG	Cornwall Seal Group
CV	Climate Vision
CWT	Cornwall Wildlife Trust
DEFRA	Department for Environment, Food & Rural Affairs (UK)
EA	Environment Agency
FL	Functional Landscape *(Narrative)*
HE	Human-Environment Interaction *(Narrative)*
NE	Natural England
NS	Natural Systems *(Narrative)*
NT	National Trust
Par'lor	Parish Councillor (in interview analysis)
R	Local Resident
SS	Surf School
SSSI	Site of Special Scientific Interest
SWCPA	South West Coast Path Association
TP	Towans Partnership
UT	Upstream Thinking Project
VC	Visit Cornwall
VB	Visual Beauty *(Narrative)*

List of Figures

List of Tables

1 Introduction

> *"Adapting to climate change is, to a large extent, a local process.*
> *Effective solutions to climate challenges should be sensitive to the*
> *local context."*
>
> Bugler & Palin (2017: 1)

The majority of scholars from the natural and social sciences agree that our planet's climate is undergoing anthropogenically accelerated change. Rising frequency of extreme weather, droughts, sea level rise as well as tidal and river flooding are altering the physical environments of many locations, and adapting to current and projected impacts of these alterations has made it to the top of most governmental agendas within and beyond Europe. Societal responses to the changing climate take shape in the form of individual action, but also new government policies and adaptation strategies addressing a multitude of societal issues from the local to the international level (CEC 2007, EEA 2013, IPCC 2013, Uk Climate Impacts Programme 2015). However, it is not only the impacts of the changing climate that alter the physical-material fabric of the spaces affected by it. Whereas some adaptation measures are structural or strategic in nature, others lead to the material reshaping of the spaces in which climate adaptation takes place (Adger et al. 2011). Although academic research into societal processes of climate change adaptation has advanced significantly over the past decades, however, our understanding of the factors that shape adaptive responses in their local and societal contexts is still limited. In order to advance understanding of the factors that shape processes of climate adaptation at the local level, this study is concerned with the individual and collective values that underlie different approaches to climate adaptation held by a particular group within society: those actors[1] who are responsible for deciding about and implementing physical-material adaptive measures on the ground. In addressing this question, this book sets particular emphasis on the role of people-place relationships and the societal constructedness of landscapes. This first chapter will lay out the broad thematic background of this study and outline the objectives and research questions that

---

[1]   I define as 'actors' quite broadly any societal groups or individuals involved in processes of decision-making and the implementation of the same in the context of climate change adaptation. ADGER ET AL (2005: 79), for instance, list as examples for such actors "individuals, firms and civil society, [...] public bodies and governments at local, regional and national scales, and international agencies".

© Springer Fachmedien Wiesbaden GmbH, part of Springer Nature 2019
V. Köpsel, *New Spaces for Climate Change*, RaumFragen: Stadt –
Region – Landschaft, https://doi.org/10.1007/978-3-658-23313-6_1

guided the theoretical, methodical, and empirical considerations. In the second half I will introduce in brief the case study area of Cornwall (UK), and give an overview of the structure of the subsequent chapters of this thesis.

1.1 Context of this Study

Many landscapes across Europe have begun to alter due to the impacts of climate change: increased wave action and coastal erosion, flooding, and extreme weather events such as storms or heavy rainfall are transforming the physical-material fabric of stretches of land (Climate UK 2012, Environment Agency 2012b, IPCC 2013). Where physical spaces are changing because of the impacts of climate change, it becomes necessary for governments and local communities to respond to these developments. The question of how to adapt both the natural and built environment as well as societal practices to the changing climate has therefore become an increasingly prominent topic both for political and private actors in many countries within and beyond Europe (CEC 2007). Consequently, adaptation strategies have been published by official institutions from the European Union, national governments and administrations as well as by non-governmental bodies on a variety of scales (see e.g. CEC 2007, Bundesregierung Deutschland 2008 Government of the United Kingdom 2013, EEA 2013). Needless to say, these documents vary in their focus areas, strategy- and practice-orientation as well as temporal scope and degree of detail (Wilson 2007, Moser & Ekstrom 2010, EEA 2013). What they have in common, however, is the recognition that climate change is a phenomenon which is both global and local, has diverse and wide-ranging impacts and brings with it alterations to our natural and built environment, infrastructure, production chains and personal lifestyles. To respond to this vast range of challenges, adaptation strategies in many cases cover a large array of sectors, actors and locations. From built environment and infrastructure to human health, agriculture, tourism and financial management, strategies to respond to the changing climate cover almost all aspects of societal life (Bundesregierung Deutschland 2008, Biesbroek et al. 2010, EEA 2013, Government of the United Kingdom 2013). However, recent literature has shown that if strategies concerning adaptation to climate change are developed by committees on a national scale they often fail to adequately address the challenges resulting from a changing climate or, respectively, those that are perceived as urgent at the local level (Adger et al. 2009, Juhola & Westerhoff 2011, Adger et al. 2013). Amundsen (2015: 262) finds that "a number of adaptation policies and plans addressing climate change impacts […] are based solely on technical solutions, while failing to take into account emotional attachment to place and the knowledge held by the people living there". Indeed, the development of adaptation strategies and related policies is often based on impact- and cost-benefit analysis and does not succeed in addressing local needs or finding acceptance among local populations (Adger et al. 2009, O'Brien & Wolf 2010, Adger et al. 2013). After all, the application of adaptive measures, even those drafted by national governments, always takes place on the ground and within local contexts (Relph 2008, Adger et al. 2013, Geoghegan & Leyshon 2014, Köpsel et al. 2016). In order to understand what factors shape the realization of adaptation measures at the local level, I argue in line with Leyshon & Walker (forthcoming) and Geoghegan & Leyshon (2014), that it is vital to shift attention to the local; to focus on the values and perceptions of those local actors who are responsible for deciding between an array of adaptation options and implementing the chosen solutions on the ground.

1.1.1 Local Embeddedness of Climate Change Adaptation

In order to understand local perceptions of and responses to a changing climate, Hulme (2008: 6) demands that "discourses about […] climate change […]be reinvented as discourses about local weather and about […] local physical objects and cultural practices". The author followed up on this demand two years later in his renowned book 'Why We Disagree About Climate Change', investigating the multi-faceted reasons for disagreement about the phenomenon of climate change between different groups and individuals in society. Hulme discusses a number of topics influencing people's perceptions of climate change: he considers the effects of historical understandings of climate on current discourses; the economic values societal groups ascribe to certain material goods; as well as the importance of distinct world views and risk perceptions for evaluating climate change (Hulme 2009). N. Adger moreover argues that not only perceptions of climate change but also societal responses to its impacts are guided by cultural values, ethics, and symbolic meanings which influence "the identification of risks, decisions about responses, and means of implementation" (Adger et al. 2013: 112) (cf. section 2.1).

To understand what shapes the decision-making about and implementation of climate adaptation measures at the local level, it is therefore necessary to view both climate change impacts as well as adaptation activities in the light of their societal and cultural contexts (Barnett 2010, EEA 2013, Adger et al. 2013). While a large number of scientific studies are concerned with "the more ecological, technical, and economic that is, physical aspects of [*adaptation*], they have neglected important psychological, symbolic, and particularly emotional aspects" (Agyeman et al. 2009: 509). Therefore, scholars from human geography and other related fields have shifted their foci from technical or economic analyses of adaptation activities towards understanding processes of climate adaption as culturally embedded within specific local and regional contexts (e.g. Adger et al. 2009, 2011 2013, Amundsen 2015, Döring & Ratter 2017). In this context, it has been increasingly recognised in human geography and related disciplines that societal adaptation to environmental changes requires public participation and close attention to the specificities of individual places (Adger et al. 2009, 2013). "The complex and dynamic nature of environmental problems", Reed (2008: 2417) argues, "requires flexible and transparent decision-making that embraces a diversity of knowledges and values".

A number of existing studies exist that scope public perceptions of climate change and attitudes towards adaptation (cf. Taylor et al. 2014). What remains under-researched to date, however, are the knowledges and values of those actors within society that are responsible for making concrete decisions about which physical-material adaptation measures should be implemented on the ground. The range of such actors stretches from local and regional government bodies to environmental management organisations, interest groups, and formal or informal partnerships focusing on specific topics at the local level. In this context, Bugler & Palin (2017: 1) underline that "[m]ulti-stakeholder, multi-level governance approaches are increasingly recognised as best practice for local adaptation" (see also Gerkensmeier & Ratter 2016). Local staff of national or regional landscape management organisations therefore play a particular role in the implementation of adaptive measures. Although bound by organisational guidelines and policies, their decisions at the same time are embedded in local framings of climate change and the places in which they live and work. Thus, local organisational staff act at the interface between expert thinking and contextualised, culturally mediated, personalised understandings of climate change and adaptation (Geoghegan & Leyshon 2014). Especially when a large number of actors are involved in the implementation of adaptation measures at the local level, it is precisely such diverging understandings of both climate change and the places affected by it that can be causes of conflict (Adger et al. 2005). One important facet of such subjective values, I argue, has to date received insufficient attention in academic research: those individual and collective perceptions and interpretations of the places affected by climate change that are held by the local actors responsible for decision-making about and the implementation of physical-material adaptation measures.

Missing 'Ground' in Adaptation Research

> *"Useful though recent work on the emerging cultures of climate change has been in identifying potentially fruitful directions for research which is more 'grounded', it is ironically the 'ground' that is missing from recent accounts."*
> (Brace & Geoghegan 2010: 287)

When physical-material climate adaptation measures are implemented on the ground, they often take up land previously used for other purposes. There is growing recognition in existing literature that adaptation will alter land use concepts and potentially create conflicts with alternative uses, for instance when restoring flood plains or creating green spaces in densely populated urban areas (Scheele & Oberdörffer 2011). Such re-shaping of spaces through human activities, be it due to climate change or other reasons, is never merely a technical process. Although human interaction with the environment is shaped by societal attitudes, perceptions and symbolic meanings, place is problematically separated out from its constitutive social relations in existing climate adaptation research (Tomaney 2015, Devine-Wright 2015). Likewise, subjective constructions of place and landscape among actors from local organisations and government, as well as the symbolic significance that specific locations hold in the context of climate adaptation, remain under-researched in existing studies (Köpsel et al. 2016).

It is not only the perceptions of and values connected with certain places and landscapes that remain understudied in the context of climate adaptation. Adger et al. (2011: 20) find that "[…] there are limits to adaptation options for places and their cultures", acknowledging that a society's relationship with their physical environment shapes the array of climate change responses taken into consideration. The societal constructedness of places influences the outcomes of adaptation processes which, in turn, may lead to a physical-material re-shaping of respective places (Köpsel et al. 2016). How in particular the relationships between people and 'their' places shape different ways of responding to developments as abstract as climate change, and what conflicts arise from contrasting understandings of places and landscapes in this context, is a question that to date remains unaddressed. It is therefore the aim of this study to understand the role of people-place relationships for decision-making processes around climate adaptation at the local level, and investigate how these processes lead to the physical-material alteration of the places affected by a changing climate. The specific focus of this study is on the roles of local actors in environmental and landscape management, and the interplay between their subjective and collective constructions of the places and landscapes they seek to adapt to a changing climate.

Grounded in the empirical data, particular emphasis will be placed both on the theoretical framework and on the discussion of results on the topics of landscape as a commons, differing understandings of the human-nature relationship, as well as the question of physical access to the landscape.

The case study area for the investigation of these foci is Cornwall in South West England and, constituting an in-depth case study within this broader context, the coastal headland of Godrevy. Before introducing the case study area in more detail, the following subchapter will present this study's objectives and research questions.

1.2 Objectives and Research Questions

Ratter (2001) (referring to O'Riordan 1971: 19) understands 'management' as culturally and societally embedded practices of decision-making and the implementation of these practices which "involves [subjective] judgement, preference, and commitment" of the actors involved "whereby certain desired [...] outcomes are sought". Whereas processes of formal planning are classically characterised by top-down approaches and the adherence to relatively static tools and strategies, management processes according to Ratter (2001) are importantly based on the participation of a variety of actors, flexibility of approaches, and the readiness for "adjustment of the trajectory" if needed (Ratter 2001, Ratter & Gee 2012: 136). At the same time, management is always a political undertaking influenced by power relations and the weighing of risks (Ratter 2001). According to Leyshon & Walker (forthcoming), 'landscape management' can be described as a "wide range of individuals and statutory and non-statutory organisations involved in the conservation of landscape and biodiversity, the design and administration of agricultural stewardship schemes, management of designated areas, and the protection of the natural environment more broadly." Leyshon & Walker focus their definition on the topics of biodiversity, agriculture, and nature protection. However, the understanding of the term in this study also includes the management of landscapes in the contexts of tourism and leisure, community development, heritage preservation, or the construction of renewable energy structures – in short, any activities that (re-)shape the physical fabric of landscapes. Such activities can be either formal planning processes or informal; can be driven by one actor or an interplay between different actors; are mostly motivated by actor-specific agendas; often lead to the physical-material reshaping of stretches of land; and can have both intentional and unintentional outcomes. In this sense, the development of outdoor activities for tourists, the reclamation of land for agricultural production, or the protection of local wildlife are here all understood as specific goals of landscape management.

All processes of landscape management hold the potential for conflict, especially when several actors with competing values and priorities are involved (Adger et al. 2005, Bugler & Palin 2017). Such diverging perspectives can exist in national strategies and the values of local staff of governments (comp. Adger et al. 2005); in the values of different actor groups involved in landscape management and climate adaptation (Köpsel et al. 2016, Leyshon & Walker forthcoming); and in the values of organisational actors and the local population (Geoghegan & Leyshon 2014, Bugler & Palin 2017). To contribute to a better understanding of the influence of diverging values on climate adaptation locally, this study examines the role of people-place relationships and perceptions of landscapes among actors responsible for the implementation of adaptive measures on the ground. I particularly address two critical points largely overlooked in current research on climate adaptation: the influence of the relationships such local actors have with the physical spaces affected by the changing climate at the local level, and the concrete material manifestations of adaptive processes in these spaces. The main objective is therefore to investigate the following overarching research question:

How are processes of climate change adaptation influenced by the social constructedness of the affected places, and how in turn does an abstract phenomenon like climate change become evident physically-materially in a socially constructed place through concrete societal actions?

The theoretical concept for addressing this objective is inherently geographic: that of landscape, from a constructivist perspective (see chapter 3). To understand what role differing perceptions of landscapes play in local climate adaptation activities and, vice versa, how physical spaces are re-shaped through respective activities, five research questions will be answered throughout this book based on empirical data from Cornwall (UK).

1. How do different local actors in landscape management subjectively perceive the landscape and landscape change in Cornwall, and how is their landscape perception informed by professional and personal knowledge and identities and their interrelationships?

This question is based on an analysis of the actors involved in climate adaptation in Cornwall, their roles and responsibilities. The focus here is on the collective landscape constructions shared among these actors, and the perceptions of and personal relationships with the landscape of single individuals in landscape management. To investigate not only perceptions of the places and landscapes but also of the changing climate and its impacts on Cornwall and Godrevy, the second research question is:

2.   How do these actors make sense of climate change, its potential or perceived
     effects on Cornwall, and its perceived impacts on the landscape?

Based on a values-based perspective on climate change and adaptation (see section 2.1), I view as vital for understanding the societal dynamics of climate adaptation to comprehend how different actors' choices of adaptive approaches are
grounded in their perceptions of the landscapes they manage. Hence, the third
research question asks:

3.   What approaches to climate change adaptation might result from the differ-
     ent perspectives on the landscape? To what extent do these approaches im-
     ply distinct relationships with and understandings of the landscape?

Building on these different perceptions of Cornwall's landscapes and resulting
adaptive approaches, the fourth question grounds the process of climate adaptation in a concrete physical space focusing on Godrevy headland by asking:

4.   To what extent do specific places or landscapes attain symbolic significance
     in debates on landscape, climate change and adaptation in Cornwall?

Lastly, the fifth research question moves away from a focus on symbolic values
and the mental constructedness of places, and puts emphasis on the physical-
material consequences of different adaptation approaches grounded in distinct
constructions of local landscapes:

5.   To what extent do processes of landscape construction and climate change
     adaption result in physical-material changes in the landscape? How can dif-
     ferent perspectives on these landscape changes be explained against the
     background of specific local debates?

Scrutinizing adaptive landscape management activities at Godrevy, this question
addresses the role of different landscape constructions in the implementation of
climate adaptation measures. Based on the analysis of diverging perceptions of
landscape and climate change I investigate how varying social constructions of
landscape result in distinct viewpoints on the future spatialities of material adaptation measures; and how these perspectives shape decision-making processes
among local actors. This study's empirical data importantly contributes to understanding multi-actor processes of climate adaptation in their local contexts, with
special focus on the social constructions of the landscapes undergoing change.

    The local context in which this study investigated adaptive processes is of
great importance for understanding its empirical results. In the next section, I
therefore introduce the research area of Cornwall (UK), putting emphasis on the
relevance of the topic of landscape and on current climate change adaptation activities. I then give an overview of the structure and contents of this book.

1.3   The Case Study: Cornwall (UK)

Cornwall is a rural peninsula in South West England stretching into the Atlantic Ocean. It is home to half a million people and governed by Cornwall Council (in the following also: the Council) (Cornwall Council 2016b). With only five towns of around 20.000 inhabitants, Cornwall is mainly characterized by villages, farms and a variety of protected landscapes (Cornwall Council 2012, 2017c). Its geography is shaped by water both from the sea and inland as "[t]he sea forms the northern, southern and western boundaries, with the River Tamar forming the eastern border with Devon and Plymouth, giving it its strong maritime character" (Cornwall Council 2016b: 9) (see Figure 1). Situated at least five hours away from London by train or car, the county can be described as remote from the geographical and political centre of England (ibid).

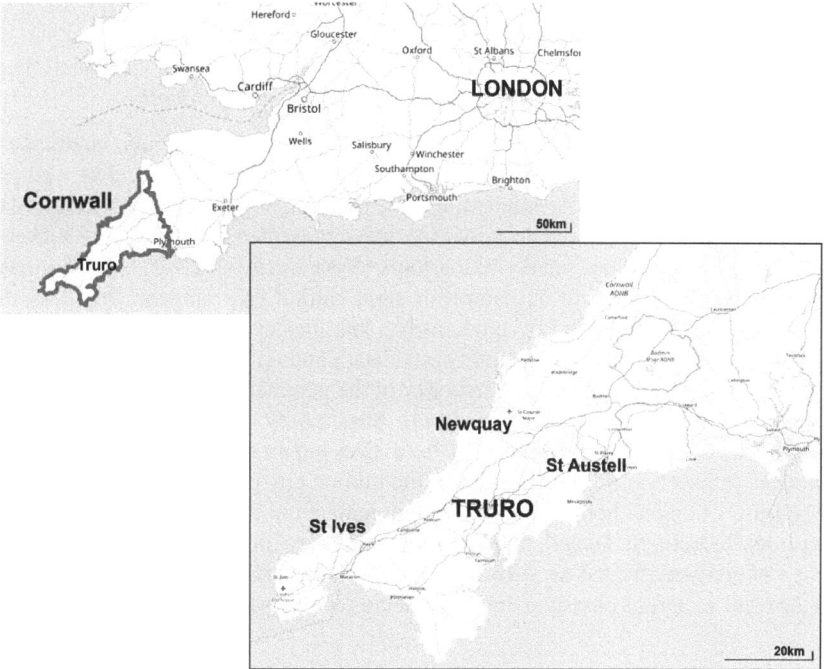

Figure 1: Geographical location of Cornwall
Source: © OpenStreetMap Contributors (www.openstreetmap.org), modified

Historically, Cornwall's economy has been renowned for tin and copper mining since the 11th century. With technical innovations exported to many parts of the world, the county was one of the motors of the British industrialization (CIoS Futures Group 2017). Other important economic sectors were fishing and small-scale agriculture. After the decline of the mining industry in the late 19th century, the tourism sector rose to the top of Cornwall's economy (Hale 2001, Cornwall Council 2016b). Once thriving financially, the county today ranks as the poorest in England and depends on EU structural funding. Cornwall's present-day economy is based on agriculture and tourism and relies heavily on its physical environment (Cornwall Council 2016b, CIoS Futures Group 2017, Cornwall Council 2017b). However, tourism as the main driver of the economy is seasonal and the local fishing sector is in decline (Trentmann 2014). Due to its economic focus on agriculture and tourism, the local landscapes of Cornwall play an important role in both its regional[2] identity and economic activity; a fact that I will expand on in the following section before introducing the activities undertaken in Cornwall with regards to climate change.

1.3.1   "Think Cornwall, Think Landscape"[3] and Regional Identity

Cornwall is known far beyond the county's borders for its spectacular landscapes consisting of small fishing villages, beaches, moorlands, and natural as well as cultural heritage sites. The county therefore ranks among the most-visited tourist destinations in the UK, well-known for and represented by pictures of bucket-and-spade tourism and scenic hiking tours (Visit Cornwall 2013, CIoS Futures Group 2017). Regarding its natural or semi-natural environment, the Cornish peninsula is characterised by modern-day and ancient pastoral farmlands, pastures, open woodland, moors, and coastal heathlands (Cornwall Council 2015a). Once substantially wooded, the majority of the peninsula's land surface is dominated by agricultural land today (ibid). Structurally, this agricultural land in Cornwall is characterised by small field sizes separated by so-called Cornish hedges (cf. section 5.1). The relicts of the mining era, primarily mine shafts and the ruins of engine houses, prominently dominate the scenery of the north coast around Redruth, St Austell, and West Penwith (see Appendix 1).

Moreover, the coast with its diverse characteristics importantly shapes Cornwall's natural environment. Connected to its Coast is a maritime cultural

---

2   With respect to the English governmental system, Cornwall is formally a ceremonial county and part of the region of South West England. However, when referring to the 'region' of Cornwall in the course of this study, I understand the term not as a political attribution, but refer to the area within the geographical boundaries of Cornwall: its coastline as well as the river Tamar separating it from its only neighbouring county of Devon.

3   CORNWALL COUNCIL 2016c

heritage in the form of a history of harbours and seafarers, fishing, and in recent days the region's popularity as a tourist destination (cf. e.g. Visit Cornwall 2013). Although a primarily rural county, Cornwall has a large number of smaller and medium-sized settlements, which often feature historic town centres. Furthermore, in recent years substantial areas of new housing development have been built around the existing settlements (see Appendix 1). When it comes to built cultural heritage, other features typical for Cornwall are its historic castles such as Tintagel Castle on the north coast, and manor houses from medieval times and the 18th and 19th century when wealthy aristocracy developed in Cornwall. The high number of town and village churches, some with Celtic origins, are thereby an indicator of the historic importance of religion in the region.

Over thirty percent of Cornwall's land area is under landscape designations such as Areas of Outstanding Natural Beauty (AONB) or Sites of Special Scientific Interest (SSSI), with long sections of its 290 miles of coastline managed by the National Trust (Cornwall Council 2016c, 2016b, National Trust 2017a). Additionally, 300 miles of the South West Coast Path stretch along Cornwall's shoreline (South West Coast Path Association 2017a). Historically, these scenic landscapes of Cornwall have been a popular subject for poets and painters especially due to their visual attractiveness and particular light conditions (St Ives Society of Artists 2017). The well-known 'Poldark' television series, has also made an important contribution to establishing Cornwall's industrial heritage from the 16th to 19th century mining era and related relict landscapes as foundational elements of Cornish identity (Beer 2016). The remains from this era, primarily in the region around Camborne and Redruth in west Cornwall, were awarded 'World Heritage' status in 2006 (UNESCO 2017, Cornish Mining World Heritage 2017).

Cornwall's Landscapes and Regional Identity
An important factor when researching societal processes in Cornwall is its strong regional identity[4] which is inseparable of its geographical location and physical environment (Tregidga 2012). Writing about sense of place on the peninsula, Marsden (2014) finds that not only its regional identity makes Cornwall a county different from the rest of the UK, but also its landscapes. "No other region of England", he argues, "offers such a range of dramatic landscapes, nor carries such a freight of mythology and projection" (ibid). It is these landscapes and "[t]he mild climate and coastline [*which*] have been attractions for many

---

4   Drawing on the work of PAASI & ZIMMERBAUER (2011: 167), I understand as regional identity the "[r]egional symbols and meanings [which] bring together the past, present and future of a region, and they [...] a key role in making the region a meaningful unit for social and economic life both 'inside' and 'outside'".

generations of British tourists escaping from the cities, seeking the picturesque" (Hale 2001: 185). In close relation to both the Cornish landscapes and the importance of tourism for the county, prevailing topics in existing literature about Cornwall are its mining heritage (e.g. Deacon 2010); medieval castles and Celticness (Emerich 2012, Giles & Cripps 2012); fishing, farming, and rurality (Deacon 2010, Keys 2012), tourism as well as surfing (Swallow 2012, Harasta 2012). Moreover, "symbols such as cliffs and rolling surf are grist to the mill of the advertising industry, inventing commodities with the magic ingredient of 'Cornishhness'" (Deacon 2010: 86/87). Therefore, whereas regional identities in many of England's counties are not particularly strong, Cornwall constitutes a case study where local politics and policy are importantly influenced by the self-perception and identities of its population (Woodcock 2015). Moreover, discussing competing heritage interpretations in the county, Hale (2001: 186) finds that "[m]any people in Cornwall do not consider themselves to be English, nor do they consider Cornwall to be a part of England". Researching sense of place in the county, Marsden (2014) argues that this specific identity is rooted in a number of circumstances. Firstly, it is owed "to the quirk of geomorphology and tectonics that placed the sea on three sides and made most of the fourth out of the river Tamar[5]", and therewith to Cornwall's distinct geographical location (cf. Deacon 2010).

Secondly, the county withstood conquest by both the Romans and the Saxons, and domination by the Tudors was fought bitterly during the 14th and 15th century. Moreover, up until the 19th century Cornish was a spoken language in the region. Today, seemingly odd place names are reminders of the originally Cornish settlements of medieval times. During the past two decades, there has been a revival of both the Cornish language and of the self-perception as being Celtic rather than Anglo-Saxon in origin among parts of the local population (ibid). It is partly owed to this renaissance of the Cornish among only a part of the county's population that "Cornwall is a contested territory; as a result of a number of economic, linguistic, religious and political features, [it] has continually asserted a sense of difference from England" (Hale 2001: 186). The Devolution Deal signed with the UK's Central Government in 2016 is the latest evidence of the still remaining wish of Cornwall's inhabitants to be more independent from London, and shows that the county is viewed as separate from the rest

---

5   The river Tamar cuts through the South West of England from its north to its south coast, separating Cornwall from Devon, its only neighbouring county. Crossing the Tamar bridge at the town of Plymouth is the only way to travel into Cornwall by car or train. The Tamar thus constitutes both the physical border of Cornwall and, at the same time, is "invested with a symbolism that is to many much deeper than merely crossing an inconvenient stretch of water"; separating Cornwall from the rest of England (DEACON 2010: 48/49, cf. MARSDEN 2014).

of England at least by a proportion of its population (Cornwall Council & Her Majesty's Government 2015).

Actors in Landscape Management in Cornwall
Landscape management in Cornwall is undertaken by a variety of actors with often very different foci and priorities. As the county's local government, Cornwall Council is actively promoting the benefits of its attractive local landscapes in the context of both tourism and regional development, and a variety of policy decisions are based on the particular narrative promoted by the Council (cf. section 5.2). The Cornwall Landscape Character Assessment – a report from 2011 dividing Cornwall into 43 areas of distinct landscape character on the basis of physical features and characteristic cultural heritage (Cornwall Council 2011a: 43). In charge of neighbourhood planning, the local Parish Councils are responsible for developing building plans for the surrounding areas, therein also considering particular features of the local landscape (Cornwall Council 2017e).

In addition to the governmental bodies mentioned above, a number of government-related and charity organisations are involved in landscape management in Cornwall. The largest of them are the National Trust and Natural England. The National Trust for Places of Historic Interest or Natural Beauty (hereafter: National Trust or the Trust) is Europe's biggest conservation organisation owning and looking after over 700miles of coastline, a quarter million hectares of land, as well as more than 500 historic properties and buildings in the UK which are deemed to be of national importance and worthy of preservation. In Cornwall, the Trust has a regional office near Bodmin. The organisation's main sources of income are donations as well as membership fees (National Trust 2016a).

A second landscape management organisation in Cornwall working at county scale and tied into a national network of policies and strategies is the Areas of Outstanding Beauty Partnership (AONB). There are twelve AONBs within Cornwall which have "the same status and level of protection as a National Park", and the Partnership advises Cornwall Council on planning matters around these areas (Cornwall AONB Partnership 2017). The AONB Partnership publishes a Management Plan every five years and works in partnership with the National Trust, Natural England, Cornwall Council, Cornwall Wildlife Trust, Rural Partnership, as well as the Cornwall Sustainable Tourism Project (CoaST).

Focused primarily on nature protection, Natural England advises the UK Government on matters to do with the natural environment. The organisation's foci lie on terrestrial and marine biodiversity, access and engagement to and with the landscape, support to the Council's planning system, as well as wildlife management (Natural England 2017a). Natural England has an area office near Truro in Cornwall. The organisation is also responsible for designating Areas of

Outstanding Natural Beauty (Natural England 2015). Another important actor in Cornwall regarding the natural environment is the Cornwall Wildlife Trust, a county-wide charity organisation which manages over fifty nature reserves and over 5,000 hectares of land in the region (Cornwall Wildlife Trust 2017b). One of the organisation's projects, 'Upstream Thinking', is particularly concerned with sustainable river catchment management, flood alleviation, and the improvement of the pollutant rates from farm runoff (Upstream Thinking 2017). In the tourism sector, Visit Cornwall as the county's tourism board and the Cornwall Sustainable Tourism Network (CoaST) are the most influential bodies in the case study area (CoaST Network 2017).

In view of the many landscape designations in Cornwall and its dependence on agriculture and tourism, the phenomenon of landscape does not only have particular relevance for local policy-making and planning, but plays an important role also in Cornwall's regional identity and economic profile (Cornwall Council 2016b). G. Tregidga (2012), Head of the Institute for Cornish Studies at Exeter University, argues that in order to understand societal processes in Cornwall, any context-sensitive research about the region needs to consider this close relationship between Cornwall as a place and its regional identity. This close relationship, as this study will go on to argue, also plays an important role in how climate change adaptation is approached in the county. To establish the context of any adaptive activities in Cornwall, the next section outlines how the peninsula is already affected by the changing climate today, and which actors are involved in adaptation processes.

1.3.2 Cornwall under a Changing Climate

As a peninsula stretching out into the ocean, "Cornwall has a mild climate, strongly influenced by the sea, the Gulf Stream and regular weather fronts from the Atlantic" (Cornwall AONB Partnership 2011b: 33). However, physical impacts of climate change are becoming visible in the Cornish landscape in the shape of coast and catchment flooding, cliff erosion, as well as heavy rainfall and storm events (Cornwall Council 2011b). Recent events of severe river and tidal flooding in the winters of 2013/14 and 2015/16 constitute examples of physical climate impacts on the region leading to property damage, road closures, as well as the temporary cut-off of the only train line connecting Cornwall to the rest of the country (Climate UK 2012, Cornish Guardian 2015). Climate change projections for South West England estimate a significant increase in such episodes, accompanied by more frequent incidents of extreme weather and a shift in rainfall patterns (Environment Agency 2012b, Climate UK 2012). Cornwall is predicted to experience up to 50% less rainfall in the summer and 10-30% more during winter by 2080. Particularly in the context of proceeding erosion, many coastal zones in the South West will face significant changes in the next two decades, for example through sediment erosion and re-deposition. Moreover, sea level rise and increasing storm frequency are expected to increase areas of coastal squeeze where assets close to the shoreline might be lost due to an eroding coast (South West Regional Assembly 2008, Cornwall AONB Partnership 2011b, Climate UK 2012). 279km of coastline protected by the National Trust is threatened from erosion and 852 hectares are in danger of being flooded, especially along the south coast, creating the need for improvement of coastal defences and relocation of structures further inland. Numerous cultural and natural heritage sites in the region are already threatened by climate change (National Trust 2008). Furthermore from a touristic perspective, a decline in visitor numbers has been observed in past years, and the summer of 2012 is referred to as the wettest in the past century (Visit Cornwall 2013).

No active intervention (NAI):	A decision not to invest in providing or maintaining defences or natural coastline.
Hold the line (HTL):	Maintain or upgrade the level of protection provided by defences or natural coastline.
Managed realignment (MR):	Manage the coastal processes to realign the 'natural' coastline configuration, either seaward or landward, in order to create a future sustainable shoreline position.
Advance the line (ATL):	Build new defences seaward of the existing defence line where significant land reclamation is considered.

Figure 2: Policy options from the Shoreline Management Plan
Source: Cornwall Council et al. 2010: 2

A number of organisations in Cornwall are concerned with the various facets of its natural environment that are affected by the changing climate. Being a national organisation but holding regional offices in the South West of England as well as Cornwall, the Environment Agency (hereafter: the EA) is responsible for flood management, river water quality, and environmental management in Cornwall. Recognizing climate change as a significant development in recent years, the EA is especially concerned with inland flood management and alleviation (e.g. Environment Agency 2012b).

As a signee of the Nottingham Declaration[6], Cornwall Council acknowledges climate change as humanly accelerated and potentially harmful to the local environment and infrastructure. A number of the Council's publications

6   The Nottingham Declaration, which was launched at a local government conference in Nottingham in 2000, is a voluntary pledge of local authorities to address the social, environmental and financial challenges of climate change (see e.g. SOUTHEND ON SEA BOROUGH COUNCIL 2014).

underline the importance of climate change and the need to address it. The Shoreline Management Plan adopted in 2010 recognizes climate change as one of the main drivers of coastal change in Cornwall and sets the frame for the management of the region's coastlines. Influenced by the Department for Environment, Food & Rural Affairs' 'Making Space for Water' strategy (Defra 2004), this shoreline plan divides coastal management in the UK into sections: no active intervention; holding the line; managed realignment; and advancing the line (Cornwall Council et al. 2010: 2) (see Figure 2). However, concrete and location-specific adaptation measures are not proposed.

Underpinning many of Council's planning and policy decisions, the Cornwall Local Plan 2010-2030 states as one of its key issues "[t]o make the most of our environment", listing as one task "increasing resilience to climate change" (Cornwall Council 2016b: 11). Details on how this resilience could be strengthened are not given. The Council's newly released Environmental Growth Strategy, moreover, acknowledges climate change as the primary driver of "the decline of nature in Cornwall" (Cornwall Council 2017b: 4). The document lists climate change and adaptation projects as goals of supporting environmental growth in Cornwall; however, without suggesting how to tackle this challenge (ibid). Although the Council underlines the need for "significant adaptation in the design and location of buildings and infrastructure", official statements about climate change are limited to mitigation and renewable energy production as a profitable income source (Cornwall Council 2015b). Moreover, austerity politics have led to the elimination of the position of a Climate Change Officer within the Council. The Environment and Planning Department considers climate change in a number of its activities like building and flood management; a cross-sectoral approach to climate adaptation, however, does not exist within Cornwall Council. Notably, an extensive report released in early 2017 by the Cornwall and Isles of Scilly Futures Group (CIoS) develops statements on all topics important for Cornwall's development in a post-Brexit world from tourism and heritage to agriculture and research; however, climate change is mentioned solely in connection with renewable energy development (CIoS Futures Group 2017).

Whereas Cornwall Council's activity around climate adaptation is limited, other organisations in the county have moved responding to climate change to the tops of their agendas. The Environment Agency (EA) is responsible by law for managing river flooding in Cornwall and has released the West and East Cornwall Catchment Flood Management Plans, which divide the region into distinct flood risk zones and suggests respective policies (Environment Agency 2012a, 2012b). The National Trust also frames climate change as a serious threat to the natural environment and properties (cf. section 6.1). The organisation is currently in the process of developing "new ways to manage our properties in the face of a changing climate", and their coastal management policy 'Shifting Shores' was a major milestone for these efforts when it was published in 2008 and, in an updated version, in 2015 (National Trust 2008, 2015). In their Position Statement on climate change, Natural England furthermore state that "[c]limate change represents the most serious long term threat to the natural environment", and that there is "urgent need to develop strategies to enable the natural environment to adapt" (Natural England 2008: 1).

From a community organisation perspective, notable initiatives around climate change in Cornwall are the Cornwall Community Flood Forum (CCFF) and Climate Vision (CV). The CCFF is "a community-led initiative supporting communities, households and businesses at risk of flooding" founded in 2012 by Parish Councils and community groups. Its main aim is to give support and capacity building around flood management and individual property protection (Cornwall Community Flood Forum 2017). Climate Vision is a regional consultancy by scholars from the University of Exeter working to "build resilience into communities, businesses and local authorities by reducing their climate related risks" (Climate Vision 2017). However, considering the lack of cross-cutting adaptive efforts by the Council as Cornwall's government, these organisations do not position themselves within an existing regional framework of climate change adaptation policy. Much rather, organisations like the Trust or Natural England are adapting the places they look after to climate change despite the lack of a governmental adaptation strategy.

In the following section, I briefly introduce the second case study area of Godrevy. More details about the actor constellation on-site and the developments regarding landscape management and climate change will follow in chapter 6.

1.3.3   Godrevy Headland

Godrevy is a coastal headland stretching into the Atlantic on Cornwall's north coast and approximately one square mile in size (see Figure 3) (Thomas & Mann 2009). It is located in St Ives Bay and the nearest village is Gwithian in the south. Meaning 'land of small farms' in English, the Cornish word Godrevy hints at a history of agricultural use (BBC Pronouncing Dictionary of British Names 1971, Cornish Language Partnership 2017). Human dwelling on the headland goes back to 1,600 BC, with some relicts dating back to 6,000 BC. Settlement can be proven since the Bronze Age (Crisp et al. 2001, South West Coast Path Association 2017b). Today, Godrevy is characterised by coastal heathland, a steep cliff coast, beaches, a dune system, and a farm in its centre. A small, uninhabited island with a lighthouse lies off its coast, which is managed by Trinity House and is said to have inspired Virginia Woolf's novel "To the Lighthouse" (Visit Cornwall 2017a).

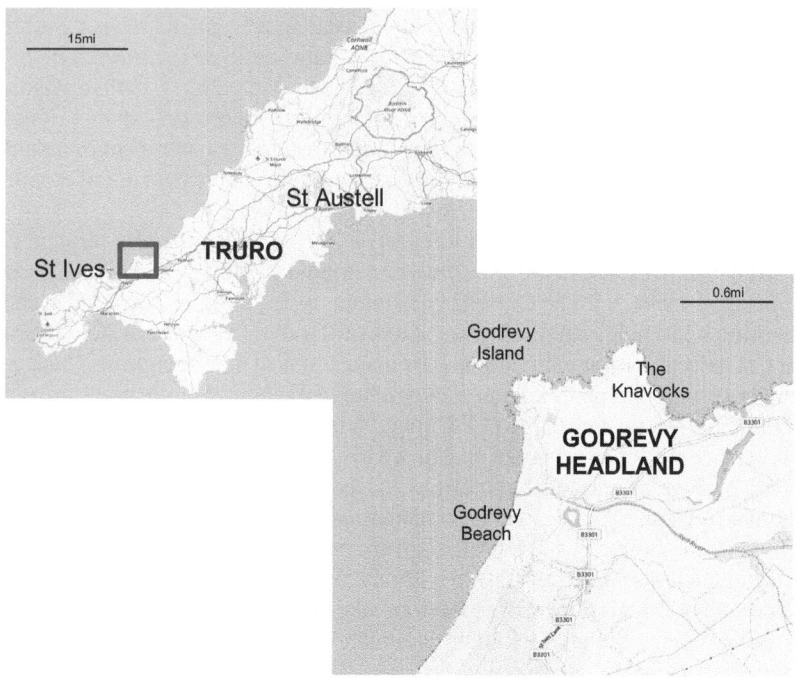

Figure 3: Geographical location of Godrevy
Source: © OpenStreetMap Contributors (www.openstreetmap.org), modified

Surrounded by a soft rock cliff coast, Godrevy is subject to ongoing coastal erosion. The rate of this erosion has been accelerating in the past decade, and the
destabilisation of the cliffs is currently causing damage to the infrastructure on
the headland such as stairs for beach access, a local road and car parks (Earlie
2015). Identified as "one of the more exposed beaches" of Cornwall, this erosion
is causing the National Trust and other local actors to be in search of an adaptive
response to the proceeding erosion (National Trust 2017e). A topic that plays a
central role in the discussions about the preferred adaptation solution is that of
climate change: Whereas actors such as the National Trust closely connect the
erosion to the changing climate and build their adaptation agenda around this
connection, other local bodies have very contrasting perspectives on the issue
(cf. chapter 6). Since these different perceptions of climate change in the context
of landscape management are a crystalliser in the local negotiations about physical adaptation measures, the Godrevy example can be viewed as a case where
the broader societal debate about climate change distinctly influences the process
of adapting to an eroding coastline at the local level.

In addition to this clear connection to the topic of climate change, Godrevy constitutes a suitable case study site for two reasons. Firstly, the headland is owned by the National Trust and is one of the most frequented outdoor visitor attractions in Cornwall (see Figure 4). It is especially popular with locals surfers, walkers, and families on beach holidays (Thomas & Mann 2009, National Trust 2017e, Visit Cornwall 2017a). Godrevy is also significant from a wildlife perspective, for example as a haul out site for Grey Atlantic Seals and a habitat for other rare species (Cornish Nature 2010). Secondly, the site is subject to a number of landscape designations with different foci and under the responsibility of different organisations. It therefore constitutes a fruitful case study site for researching the societal constructedness of the Cornish landscape. In addition to the National Trust, the AONB Partnership, Natural England, the Cornwall Wildlife Trust and a number of other initiatives are involved in landscape management at Godrevy (cf. section 6.1). The site is thus an interesting example of landscape management in a multi-actor setting. Despite its value as a case study site and its importance for local identity and tourism, however, very little research exists about Godrevy aside from geological and archaeological studies (e.g. Crisp et al. 2001); making it a particularly worthwhile case for this study.

Considering the prominent role that the issue of landscape has in the county as well as its exposure to current and projected climate change impacts, Cornwall serves as an interesting example for investigating this study's research questions. Within this context, Godrevy headland encapsulates the importance of agriculture and tourism throughout Cornwall as well as the significance of its coastline and maritime location. Moreover, it is an early example of conflict between different landscape organisations and local actors regarding the question of how to adapt to climate change and, in particular, coastal erosion. The larger case study region of Cornwall along with the smaller-scale example of Godrevy thus constitute useful locations for researching the impact of distinct people-place relationships on the societal process of adapting to a changing climate locally.

Figure 4: Godrevy Beach, cliffs, life guard hut, and lighthouse
Source: own photo, 2016

1.4 Structure of this Book

This book is divided into 8 chapters. In chapter 2, I review the contemporary literature that forms the context of this study: values-based climate change adaptation research, and particularly the role of people-place relationships and place attachment for processes of climate adaptation. I then summarise existing academic understandings of 'experts' and 'locals', and argue for a need to transcend this dichotomy to gain a better understanding decision-making processes around climate adaptation at the local level.

Chapter 3 is dedicated to the theoretical perspective upon which this study is built, namely that of social-constructivist landscape research. I firstly present contemporary academic perspectives on the concept of landscape and discuss the tensions between the concepts of landscape and place. I then link the idea of landscape as a common good with different perspectives on the normative question of physical access to it for a variety of publics. Subsequently, I develop the analytical framework of this study by expanding on the different dimensions of societal landscape construction and the ways in which landscapes are negotiated between different actor groups in society.

Chapter 4 sets out the methodology of this study and its research design. After briefly presenting the qualitative approach applied here, I focus on case study research and the use of the concept of narratives for analyzing empirical data. In the following, I introduce the qualitative methods of document analysis, semi-structured and walking interviews, and outline how these were applied during the empirical phases in Cornwall. The second half of this chapter details how the collected data were analysed using a grounded theory approach and narrative analysis.

Chapters 5 and 6 present the empirical findings of this study. Chapter 5 is focused on the societal constructions of the Cornish landscapes identified from the first set of interviews in 2015 with actors in environmental and landscape management in the mid-Cornwall area. I firstly highlight what commonly shared constructions there are of Cornwall's landscapes and landscape change. I then outline what I term the official policy narrative promoted by Cornwall Council, before focusing on four co-existing landscape narratives that provide a counterpoint to this official narrative. Moreover, I expand on the ways in which the interviewees relate to the landscape individually. The latter part of Chapter 5 is dedicated to the implications of the different narratives for adaptation to climate change in Cornwall.

Chapter 6 focuses on the case study of Godrevy. Based on empirical data from 2016, I present how processes of climate adaptation are embedded in their societal context, and how distinct constructions of the landscape influence the discussions among the local actors. Firstly, I give an overview of the actor constellation at Godrevy headland. I then focus on how the landscape narratives identified in chapter 5 influence these actors' perceptions of Godrevy, and how they perceive the impacts of climate change. In the following, I present a number of practical considerations that shape the adaptive activities at Godrevy: the question of who should have access to the landscape as well as local structures of land ownership and management. Lastly, I put into concrete terms the implications of the landscape narratives for the relocation activities at Godrevy, presenting five distinct suggestions for a physical reshaping of the landscape to adapt to an eroding coastline.

In chapter 7, I discuss the empirical findings of this study against existing literature and this book's theoretical framework. Specific focus lies on the comparison of the narratives and their implications for climate adaptation, and on the roles of local actors as individual decision-makers at the interface between a wider context and local embeddedness. The discussion of the Godrevy case emphasises the relevance of the narratives in local debates about adaptation to coastal erosion; here viewing local adaptation processes as negotiations between the actors involved in decision-making. I synthesise my findings in a Place-Based Framework for Local Climate Adaptation; a framework that introduces five 'filters' through which broader adaptation strategies translate into local contexts. Lastly, I critically review the choice of theory, methods, and the research process.

The conclusions in chapter 8 summarise the findings of this study. They outline its empirical, theoretical, and methodological contributions, and the practical implications for local processes of climate adaptation and landscape management. At the end of this chapter, I propose avenues for future research based on this book's conclusions.

2 Climate Adaptation, People-Place Relationships, and Local Actors: Reviewing the Literature

In times of a changing climate, multifaceted interactions between socio-economic and environmental processes such as storms, flooding, coastal erosion lead to far-reaching changes to specific places and spaces (Agyeman et al. 2009, Climate UK 2012, Adger et al. 2013, Köpsel et al. 2016). The impacts of climate and environmental change are situated in place and thus require adaptive responses which are sensitive to the diversity and specificity of local places (Devine-Wright 2015). Such responses to climate change are mostly divided into two broad categories: mitigation activities (the reduction of the release of greenhouse gases into the atmosphere) and adaptation activities (Wilbanks et al. 2007), of which the latter are the focus of this study. The term 'adaptation' was first used prominently by Darwin in the 19th century to describe how species adjust to their surroundings through changed behaviour, and entered the vocabulary of geographers during the 1950s and '60s in research on how societal groups made use of the resources their natural environment provided them (Orlove 2009).

Today, and in relation to climate change, in the frequently cited definition from McCarthy et al. (2001: 982) adaptation is understood as an "adjustment in natural or human systems in response to actual or expected climate stimuli or their effects, which moderates harm or exploits beneficial opportunities". As concrete examples of adaptive activities, Adger et al. (2009: 1) list the relocation of homes or settlements, changes in cropping and food consumption, the compensation of economic damage, and coping with changes to places or wildlife. Although many governments and communities recognize that the changing climate will pose new challenges that necessitate adaptation, it has until recently been somewhat neglected in the discourse about climate change. However, since the early 2000s, the body of literature on climate adaptation has grown significantly. Academics from both the natural and social sciences have taken up research on the driving forces and consequences of climate adaptation. Key studies within the field are the works of Adger and colleagues (e.g. 2009, 2011, 2013), Hulme (2009), O'Brien & Wolf (2010) and Wolf et al. (2013), and the reports by the IPCC Working Group II published every seven years since 2001 (McCarthy et al. 2001, Parry et al. 2007, Field et al. 2014). Up until recently, "much of the debate about responses to climate change focuse[d] on assessments and measures of the costs of goods, services, and technologies to reduce future impacts" (O'Brien & Wolf 2010: 232).

© Springer Fachmedien Wiesbaden GmbH, part of Springer Nature 2019
V. Köpsel, *New Spaces for Climate Change*, RaumFragen: Stadt –
Region – Landschaft, https://doi.org/10.1007/978-3-658-23313-6_2

According to Pelling (2010: 13), the four main questions in the current debate surrounding adaptation are: What to adapt to? Who or what adapts? How does adaptation occur? What are the limits to adaptation? Whereas in the global south the focus is predominantly on climate change impacts on rural livelihoods and respective adaptive responses (e.g. Bryan et al. 2013, Butler et al. 2014), the majority of European case studies concentrate on adaptation in agriculture and forestry (e.g. Lindnera et al. 2014); reconstructions of past adaptation processes (e.g. Fazey et al. 2016); urban responses to climate change (e.g. Demuzere et al. 2014); and the comparison of national and regional adaptation strategies and modes of governance (e.g. Biesbroek et al. 2010). Existing literature on the societal and individual values and beliefs that underlie decision-making around climate adaptation, however, is still relatively small and fragmented. Throughout the next sections, I review two bodies of literature that frame the research gaps that are addressed in this study. Firstly I review work on a values-based approach to climate adaptation research (2.1), and specifically the relationships that people have with the places where they live and that are altered by a changing climate (2.2). For a later reflection of the individual roles of the local actors I interviewed, I focus in section 2.3 on the current understanding within academia of 'experts' and 'lay people', and their roles in decision-making processes. Lastly, section 2.4 gives a brief summary of the literature review.

2.1   A Values-Based Approach to Adaptation Research

During the 2000s and until recently, social scientific and human geographic adaptation research had two main foci. On the one hand, an outcome-based approach was focused on how to reduce the impacts of climate change on human society, infrastructure, settlements and livelihoods (O'Brien & Wolf 2010). It centered around sectoral adjustments and adaptation strategies, was often based on model predictions, and was mainly used in planning (e.g. Bours et al. 2013). A contextual approach, on the other hand, investigated the broader social and political factors underlying vulnerability to climate change (e.g. Keller & Adger 2000). Both approaches, however, failed to take on a more subjective perspective and hence failed to understand the local and individual values shaping processes of climate adaptation. What is at stake for different groups within a society as well as the "questions of why climate change matters, to whom, who wins, who loses, and whose values count" have often remained unaddressed (O'Brien & Wolf 2010: 235). Although the recognition of the importance of individual values for understanding climate adaptation is growing within academia, research in this field is still under-represented even in the IPCC reports focusing on the human responses to climate change[7]. Therefore, O'Brien & Wolf (2010) demand in line with Adger et al. (2013) and Amundsen (2015) an integration of individual and subjective meaning in climate adaptation research. Amundsen (2015: 258) argues that to date, "a consideration of the 'subjective' dimensions of adaptation, such as values, culture and sense of place, is largely underrepresented within scientific debates on climate change".

As a response to this lack of consideration of local contexts and societal values, a new strand has developed within the field of climate adaptation research. This so-called values-based approach "draws attention to the subjective dimensions, taking into account how climate change – and responses to climate change – may influence things that are differentially valued across individuals, groups, or societies" (O'Brien & Wolf 2010: 238). As 'values', O'Brien & Wolf (2010: 233) understand the "interests, pleasures, likes, preferences, moral obligations, desires, wants, goals, needs, aversions, and attractions" that people hold towards specific issues or places, and which can differ widely between groups of a society. As a consequence, different groups and individuals perceive and evaluate climate change and its impacts differently on the basis of these values (O'Brien & Wolf 2010, Adger et al. 2013, Wolf et al. 2013). As such, Geoghegan & Leyshon (2012: 57) find that climate change is "at once a reality, an agenda, a problem, a context, a narrative and a discourse". A values-based perspective on climate change and adaptation puts the focus on the values and beliefs that different groups within society hold and asks which facets of their cultural practice and natural environment they seek to preserve under a changing climate (O'Brien & Wolf 2010, Adger et al. 2013, Wolf et al. 2013).

In a similar context, Arbuckle et al. (2015) also find that societal perceptions of climate change itself are influenced by people's beliefs about its existence and the characteristics of its impacts, as well as the societal values associated with these impacts. Moreover, the authors argue that, "individuals tend to base judgments and decisions on associative and affective models of cognition that are evolutionarily older and rooted more in feeling than in scientific facts" (ibid: 207). Especially if a phenomenon is a newly emerging topic within society, as is the case with climate change, people's attitudes are likely to be much more influenced by values, worldviews, and emotions than by statistical and abstract data (Weber & Stern 2011). The importance of a values-based research approach for understanding the dynamics of societal responses to climate change is underlined by Fresque-Baxter & Armitage (2012) who conclude that that effectiveness of adaptation interventions relies heavily on the societal acceptance of the same.

---

7   As the IPCC reports compile existing scholarship on climate change, this lack of focus on individual and place-based values mirrors the general under-representation of the topic in existing academic debates.

It is thus vital for successful adaptation planning to take into account what different groups value and are attached to. O'Brien & Wolf (2010: 233) argue therefore that "successful adaptation depends on what people perceive to be worth preserving and achieving, including their culture and identity". Existing studies from the field of values-based adaptation research show that the values underlying a societal, group or individual perspective on climate change strongly influence what their goals for adaptation are, what they wish to protect and what adaptation measures they choose to implement (O'Brien & Wolf 2010, Adger et al. 2011, Wolf et al. 2013, Adger et al. 2013, Amundsen 2015). Nonetheless, Amundsen (2015: 258), in line with O'Brien & Wolf (2010) and Wolf et al. (2013), calls for an even more contextual approach to adaptation research as "individuals and communities' responses to climate change are based on what they value, their history and their attachment to particular places". I thematise these people-place relationships in the next section with regards to their relevance for local management actors' perceptions of climate change and adaptation to its impacts.

2.2 People-Place Relationships and Climate Change Adaptation

For the past three decades scholars from human geography and the social sciences have been researching the relationships between people's identities, their attachments to places and societal responses to environmental change (Devine-Wright 2013). Furthermore, when researching the perception people have of climate change and the ways in which they adapt to it, one important facet of subjective values and meanings is the relationships people have with the places where they live (Relph 2008, Adger et al. 2011, Fresque-Baxter & Armitage 2012, Devine-Wright 2014, Amundsen 2015). Within and beyond the field of human geography, places are most commonly understood as geographical locations with their distinct physical characteristics and, more importantly, the set of societal and individual values, beliefs, and attachments superimposed on them (Tuan 1977, Adger et al. 2009, Devine-Wright 2014). In place research, which developed in the 1970s from a critique of the too rational concept of space, the focus of 'place' lies on emotional attachments to and individual perceptions of smaller-scale locations such as neighborhoods or single buildings and landmarks. Even though marginally taken up in planning and architecture, place research originated, and is still mostly referred to in human geography (see e.g. Tuan 1974). The concept of place is inherently anthropocentric and centers on human activity and meaning-making (Wattchow 2013). Whereas "[l]ocation refers to an absolute point in space with a specific set of coordinates and measureable distances from other locations", Cresswell (2009: 169) argues that "[p]lace is a meaningful site that combines location, locale, and a sense of place". 'Location' thereby determines where a place is set geographically; 'locale' refers to the material setting of a place with its "buildings, streets, parks, and other visible and tangible aspects of place"; and 'sense of place' are all those meanings, symbols, feelings, and emotions associated with a certain locale by people (ibid: 169). In the words of Tuan (1977: 6), "[w]hat begins as undifferentiated space becomes place as we get to know it better and endow it with value".

To the local population of areas affected by the impacts of climate change, such places are lived spaces of everyday life, valued for cultural heritage and personal attachments (Brace & Geoghegan 2010, Adger et al. 2011, Ratter & Gee 2012, Devine-Wright 2014). In the course of focusing on individual and contextualised understandings of climate change and adaptation, researchers have thus recognized in recent years the importance of investigating how people make sense of the changing climate and respond to it within the settings of their everyday lives, "locally, in particular places" (Relph 2008: 31). Related work from the field of human geography concentrates on the cultural dimensions of climate change and assesses how a changing climate interferes with cultural values and livelihoods, studies the role of place loss for different communities (Adger et al. 2013), researches the societal values underlying climate change perception and adaptation (O'Brien & Wolf 2010, Wolf et al. 2013), and investigates how increased levels of place attachment lead to higher willingness to engage in pro-environmental behaviour (Stedman 2002, Fresque-Baxter & Armitage 2012, Devine-Wright 2014). Changes to places, Devine-Wright (2014) argues, are often correlated with negative impacts on place attachment and people's place-based identities. Place change can also have positive impacts, however, and the question remains open as to whether strong attachment to places and landscapes rather hinders or supports the acceptance of place change through responses to climate change. Focusing especially on the social limits to climate adaptation, Adger et al. (2009: 346) emphasize the importance of investigating people-place relationships since "adaptation activities are inherently local and are necessarily based on contextual knowledge", and thus deeply rooted in the life worlds of individuals and communities in their local places. Researching the connections between sense of place and societal responses to environmental challenges, Relph (2008) moreover argues that recognizing people's attachments to places and the importance of place attachment for local identities is vital for finding and understanding locally embedded ways to cope with climate change.

Physical changes to places, whether environmentally or societally induced, can have significant implications for people's emotional bonds with respective locations. Regarding climate change, Devine-Wright (2014) distinguishes between two different categories of changes to places: On the one hand, places can be altered through direct physical manifestations of climate change impacts such as extreme weather events or rising sea levels. On the other hand, human responses such as the increase in renewable energy infrastructure or physical-material adaptation measures can alter the fabric of places. In the existing literature, such physical changes to places through environmental or human impacts are often correlated with negative changes to people's place attachments (Adger et al. 2011, Devine-Wright 2014). In addition to the physical changes to places discussed above, climate change "may disrupt a sense of continuity with the past […]. It may also lead to disruption to place related distinctiveness, if a place with a historic pattern of microclimatic influences becomes changed" (Devine-Wright 2014: 172). In their paper on responses to climate change and place identity, Fresque-Baxter & Armitage (2012: 253) find that it is vital to also focus on people-place relationships when investigating the role of subjective factors in adaptation. Furthermore, the authors argue that an increased understanding of how people make decisions on climate change adaptation based on their place-based values can raise the efficiency and local acceptance of adaptive measures. "People may have access to objective assets of adaptation", they conclude, "however, if the change/action is not congruent with values or overall identity, action may not take place" (ibid: 257). For Agyeman et al. (2009), it is the consideration of these attachments to places which is lacking in the design and application of policy decisions around climate change, often resulting in a feeling of neglect among the local population. After all, "[r]esponses to climate change are inextricably linked to the local context in which they emerge" (Bugler & Palin 2017: 1).

The studies introduced above jointly acknowledge the need to consider both individual and societal values in processes of environmental management and decision-making around climate change. "Elucidating place meaning", as Fresque-Baxter & Armitage (2012: 260) highlight, "can help to explore how people feel about certain types of activities such as development or infrastructure change, particularly when these activities may impact held meanings". This is especially important when adaptation activities carried out by one group have the potential to impact negatively on what another group values (O'Brien 2009). In the Norwegian context, Amundsen (2015) finds that concentrating on people's motivations to adapt to a climate change based on their place attachments is more helpful for understanding the societal dynamics of adaptation than merely studying its physical impacts. Thus, existing research into the role of place attachments and emotional bonds with places is an important basis for understanding the influence that people-place relationships have on societal responses to climate change. While many insights from the rich body of place attachment literature have already travelled into the realm of climate adaptation research (e.g. Devine-Wright 2014, Amundsen 2015), three perspectives are under-developed but, I argue, are also crucial for understanding the societal dynamics of climate adaptation on the ground. Firstly, existing place attachment research often overlooks what happens when different attachments come into correspondence with each other. Secondly, even though the attachments and meanings people associate with places are at the core of this field, place is problematically separated out from its constitutive social relations (Tomaney 2015, Devine-Wright 2015). Lastly, although the local actors responsible for decision-making and implementation of physical adaptation measures play crucial roles in reshaping spaces and landscapes at the local level, their professional perceptions of and personal attachments to the places they manage rarely stand in the focus of research. What is thus lacking from the literature are empirical studies investigating how adaptation activities contribute to the societal construction of space and place from a social constructivist viewpoint: How is place-making through climate change adaptation embedded in local and cultural contexts? How does it change the relationship between a local population and their environment? And through what processes does this relationship turn into concrete physical-material manifestations?

Whereas many existing studies focus on asking 'How do people feel about this place?', I therefore argue that a vital question for understanding local climate adaptation processes is: Which different perspectives exist on the question 'What is this place?'. A decisive role in this context, Geoghegan & Leyshon (2014) argue in one of their Cornwall case studies, is played by governance and landscape management actors at the local level. After all, it is their individual understandings of both local places as well as climate change which shape the decisions about what adaptation measures are implemented on the ground. Nonetheless, subjective understandings of place, landscape, and climate change among respective actors remain under-researched in work on societal adaptation processes (Geoghegan & Leyshon 2014, Köpsel et al. 2016). The role which such local actors take on in climate adaptation activities as well as this study's distinct perspective on their position within society is the subjects on the following sections.

2.3 Local Actors in Climate Change Adaptation

> *"The increasing attention to adaptation to climate change has not come*
> *with sufficient emphasis on the local nature of climate adaptation."*
> *(Agrawal et al. 2009: 4)*

In the context of more participatory, bottom-up forms of landscape management and adaptation planning (cf. section 1.1) and a focus within adaptation research on subjective, individual perspectives on climate adaptation (cf. section 2.1), it becomes necessary to move the actors into focus who are responsible for deciding about and implementing adaptation activities on the ground. I argue in line with the above quote from Agrawal et al. (2009) that the consideration of local contexts as well as the roles of local actors are decisive for understanding how different groups within a society adapt to the impacts of a changing climate, and how larger-scale adaptation strategies are translated and implemented on the ground. To argue why this study focuses on these actors, the following section reviews existing literature about their importance for processes of climate adaptation on the ground (2.3.1). Subsequently, section 2.3.2 discusses the distinction that is often made between so-called experts and lay people, and highlights the perspective this study takes with regard to the expert-lay dichotomy.

2.3.1 The Role of Local Actors in Adaptive Action on the Ground

Agrawal (2010) argues that "[a]daptation to climate change is highly local, and its effectiveness depends on local and extralocal institutions" (ibid: 173). When using the term institutions, Agrawal et al. (2009) refer to actors in rural settings such as "local governments, farmer's groups, community-based organisations, local associations and NGOs" (ibid: 4). Thus, the local institutions Agrawal (2009, 2010) researches are in practical terms equivalent to what I term 'local actors' in the context of this study (see Footnote 1, p. 1): those orgnisations and individuals who act at the interface between national adaptation strategies and policies, and local, situated contexts. It is especially these organisations and individuals who function as "the fundamental mediating mechanisms" for the implementation of adaptation activities in local places on the ground (Agrawal 2010: 174). Agrawal et al. (2009) underline the "critical role" that such actors "play in determining the nature and outcomes of adaptation strategies" (ibid: 5) by ascribing to them the significant tasks of

a.   providing capacity building and training to their local staff as well as the local population regarding responses to climate change impacts;
b.   undertaking measures locally to mitigate or adapt to physical impacts such as "floods, climate related hazards, [...] migration, etc." (ibid: 4);
c.   featuring flexibility to change in their adaptation strategies;
d.   as well as acting as the linkage between higher and lower levels of government (in the case of England, e.g. Central Government and Parishes).

Uphoff & Buck (2006) differentiate between three broad categories of such local actors: public bodies (local governments and government-related agencies); civic institutions (e.g. formalized membership organisations and local community associations); as well as private, market-driven actors (e.g. private businesses or charities). Researching how strategies for adaptation to climate change and coastal erosion are implemented by local staff of the National Trust in Cornwall (UK), Geoghegan & Leyshon (2014) find that "uncertain, shifting and unexpected alliances […] form and dissolve between [local] actors in the landscape" and underline "the importance of social networks to these alliances and to the environmental visions that they promote or resist" (ibid: 644). Despite the importance of their insights, the authors' study is relatively alone in its focus on the role of local decision-makers in processes of climate adaptation. Agrawal et al. (2009) rightfully make the criticism that most existing studies in this field either focus on technical solutions, the larger-scale governance structures in which adaptation activities are embedded, or the bottom-up responses by local populations themselves. What is still scarce in existing literature is a focus on those actors who are situated at the conflicting area between the large scale and local contexts – the staff of local organisations and associations – as well as empirical case studies shedding light on the processes which shape adaptive decisions on the ground (ibid).

Furthermore, when it comes to researching distinct perceptions of the places affected by the changing climate and their implications for management processes, a focus on local professionals and decision-makers is scarce. In his 2013 book chapter about perceptual lenses in landscape studies, Howard (2013: 49) points out that "there are many professional users of the landscape where an investigation into their perceptions might be rewarding". Furthermore, the author states that "the clash between experts and locals […] is another field where there is much to be done" (ibid: 49).

As this study focuses on the actors situated between national and local contexts, I consider it important to take a closer look at how the role of these actors, often referred to as 'experts' in their fields, is conceptualized in existing literature more generally. Therefore, the following section reviews the current academic debate about the role of experts, locals, and their respective stocks of knowledge, and proposes an alternative perspective on this commonly made distinction.

2.3.2   Experts or Locals? Transcending the Dichotomy

The idea of differences between expert and local knowledge or (depending on the context) scientific and lay knowledge is not new within the social sciences and human geography. Collins & Evans (2002) understand as 'experts' individuals within a society who hold knowledge about specific topics which exceeds that of so-called lay people (ibid). With this what the authors call a "realist approach" to conceptualizing the role of experts, they criticize a more radical constructivist viewpoint on expert knowledge from the 1970s and 1980s which rejected entirely any separation between the knowledge of experts and lay people (ibid: 239). Also in contemporary literature, scientific or expert knowledge is mostly associated with "factual, objective and real knowledge", whereas the viewpoints of lay people are often displayed as "sentimental, emotional and intellectually vacuous" (Wynne 2001). There are two prevalent research foci on this topic area. On the one hand, a large amount of work within and beyond the field of human geography investigates the perspectives of *either* professionals *or* locals or lay people (Adger et al. 2011, Baptiste 2013). Howard (2013: 48), for example, demands more in-depth research into "the clash between experts and locals" when it comes to the perceptions of landscapes. On the other, many studies exist which call for a better integration of local knowledge into policy- and decision-making around climate change and adaptation (Jennings 2009, Raygorodetsky 2011). Both these bodies of literature do, however, reproduce the long-standing divide between the two distinct types of expert/scientific and local/lay knowledge.

A more extensive critique of the distinction between scientific and lay knowledge is presented by Harvey (2007: 31) who argues that "criteria such as activities and practices, knowledge domains, or sets of norms, standards and culture" often cannot be clearly attributed to one or the other. This observation was also made by Leyshon and colleagues in the course of investigating local understandings of climate change among practitioners in environmental management (Geoghegan & Leyshon 2012, Leyshon & Walker forthcoming). I argue and will show throughout this study that this observation holds equally true in the context of climate adaptation. In dialogue with farmers on the Lizard Peninsula in Cornwall (UK), Geoghegan & Leyshon (2012: 57) found that "the boundaries around lay, technical and experts knowledges are blurred" when it comes to local climate change perceptions. Likewise, it is a key argument of this study that many staff members of regional or local environmental management organisations unite both types of knowledge, professional and personal, expert and lay, and thus take on a particular role in decision-making and implementation at the local level. Thus, the boundaries between expert, professional knowledge and personal attachments and experiences are blurred in the case of actors in landscape management who work and live in the places they manage, as demonstrated in this case study. In dialogue with Cornish farmers, Geoghegan & Leyshon (2012: 62) refer to this amalgamation of different knowledges as "expert, professional knowledge […] with an embodied, visceral understanding that emerges from" living in and experiencing the place managed with regard to climate change. This implies, in particular, that the ways in which local actors understand and perceive the landscape may be interpreted in terms of the interweaving of both expert technical and lay experiential, place-based knowledge.

In agreement with Geoghegan & Leyshon (2012: 58), I argue that the constructions of landscape and climate change of professionals in landscape management are just as much "a 'mingling' of place, personal history, daily life, culture and values" as the authors found in their study of farmers experiencing the changing climate in their daily agricultural work on the Lizard peninsula in Cornwall (UK). However, "emotionality is an important quality here which is often avoided in the more rational statutory decision-making processes" (Ratter & Gee 2012: 136). As valuable as the distinction between expert and lay knowledge may be in the contexts of certain research questions, it is particularly these *hybrid* perceptions of change to places and landscapes, these melding of personal experiences and organisational policies which play a vital role in local decision-making around climate change adaptation.

Although applying an overall social-constructivist approach to climate change adaptation research, it is not the aim of this subchapter to doubt that professionals in, for instance, environmental management possess certain knowledge that differs from that of persons not working in this field. What this perspective challenges, however, is an understanding of experts as really just that: professionals without any personal involvement in the issues they work on.

2.4 Summarising the Research Gap

Grounding climate change adaptation in its local context means asking how adaptation processes are influenced by collective and individual perceptions of climate change, place and people-place relationships. It means investigating the relationship between the cultural embeddedness of place perceptions and the ways in which the abstract, statistical, intangible phenomenon of climate change is understood by different actors through the lenses of societal beliefs and values (O'Brien & Wolf 2010, Geoghegan & Leyshon 2012). Studies from human geography and related fields so far focus on general cultural beliefs which shape people's perceptions of climate change (e.g. O'Brien & Wolf 2010), the ways in which place attachment influences adaptive processes (e.g. Devine-Wright 2014), as well as how people-place relationship are altered by the impacts of climate change (e.g. Adger et al. 2011). What is lacking from existing literature, I argue, is attention to the following facets of values-based adaptation and place research:

- The societal constructedness of places, and how constructions differ between groups and individuals within society;
- the ways in which such differing place attachments and place constructions influence approaches to climate adaptation;

- a focus on those actors at the local level who make decisions about adaptive activities and their implementation;
- a differentiated investigation of the roles of such actors as situated between 'expert' and 'lay' perspectives in respective local contexts.

To comprehend how the places affected by climate change are socially constructed, and how these constructions influence local actors' adaptation activities, I employ a concept rooted deeply in human geography and having a renaissance in the past two decades: the concept of landscape, from a social-constructivist perspective. The following, theoretically oriented chapter begins with an overview of existing work from human geography and related disciplines on the topics of landscape and climate change adaptation. Subsequently, I introduce the perspective on landscape applied throughout this study, connect it with a number of related topics relevant for analysing the empirical data, and present the concept of constructivist landscape research as a useful framework for comprehending societal processes of climate adaptation at the local level.

3 The Societal Construction of Landscapes in the Context of Climate Change Adaptation[8]

> *"Clearly we are dealing with something complicated, and*
> *we cannot even say that it is a complicated object because,*
> *as many theorists have pointed out, landscape is something*
> *which is mental as well as physical."*
> *(Thompson et al. 2013: 1)*

A valuable concept that connects people, places, and identities is that of land-scape. In human geography, a small but growing body of literature has begun to analyse social processes of climate adaptation through a 'landscape lens', thus shifting the focus of research to the local level and the social, place-based values shaping adaptive activities. Scherr et al. find that in recent years "the term [land-scape], and the management and policy approaches underlying it, are beginning to gain prominence as the limits of sectoral approaches become more apparent in our interconnected, crowded, resource-constrained and climate-chaotic world" (ibid: 1). When focusing on the societal processes that shape climate adaptation, "[t]he […] framework of landscapes can provide insights into why different groups clash over the meaning of a change in the natural environment and the definition of potential human consequences from that change" (Greider & Gar-kovich 1994: 13/14). Similarly, the European Research Foundation (2010: 2) finds that a focus on landscapes "can help with many of the challenges facing 21st century society. These challenges include urban and rural transformation, […] demographic and lifestyle changes and the human contributions and re-sponses to climate change, including […] the new […] landscapes that will emerge". Outside of the field of climate adaptation research, Alkon (2004) and Soliva (2007) show that differing perceptions of or dominant stories about a land-scape strongly influence its management and governance (cf. section 4.1.2). De-spite its usefulness for addressing the relationships between people and the places they live in, research on processes of climate adaptation through a 'landscape lens' remains underdeveloped (Köpsel et al. 2016). However, few studies do

---

8   GAILING (2012b), with reference to GIDDENS' (1988) double-hermeneutics in interpretative research, distinguishes between intellectual constructions of landscape in academia and con-structions within society, but outside the academic realm (cf. section 3.1). The author refers to the latter as "weitergehende gesellschaftliche Konstruktionen" (GAILING 2012b: 149), here translated into English as 'societal constructions'. As opposed to the term 'social' (in the sense of the German word 'sozial'), I deliberately use 'societal' with reference to different actor groups within a society (in the sense of the German word 'gesellschaftlich').

© Springer Fachmedien Wiesbaden GmbH, part of Springer Nature 2019
V. Köpsel, *New Spaces for Climate Change*, RaumFragen: Stadt –
Region – Landschaft, https://doi.org/10.1007/978-3-658-23313-6_3

exist. An important contribution in this context is Brace & Geoghegan's (2010) call to investigate how everyday practices in the landscape influence local perceptions of the global phenomenon of climate change. The authors demand more in-depth climate change research that is "grounded and localized through the concept of familiar–embodied, practised and lived–landscapes of everyday life" and reason that focusing on the material manifestations of a changing climate in the landscape draws attention to the physical spaces in which climate change takes place (ibid: 296). They conclude that a landscape approach "becomes a possible means with which to organize the immediate and future, spatially and temporally intimate relations between people, flora, fauna, topography, environment and, crucially, weather" (ibid: 289). The concept of landscape links time and space, sets a focus on the physical and discursive constitution of places, and on notions of familiarity and attachment. Similarly, Bohnet & Beilin (2015: 188) argue that "the 'landscape' metaphor is intrinsically about people in places at a time when the Earth is frequently conceived as global and placeless", thus moving the focus of research to the local and taking heed of the context. In this light, the ability of the landscape lens to reveal the local embeddedness of climate change meanings has been underestimated (Brace & Geoghegan 2010). Looking through such landscape lens, I argue, is a useful means for addressing local processes of climate adaptation and grounding them in distinct perceptions of the physical spaces in which climate change takes place.

This study advances human geographic research on climate change adaptation by bringing the individual and collective social constructedness of landscapes into the discussion. At the same time, it investigates the relevance of people-place relationships for the management of under climate change and addresses the topic of landscape as a common good and the question of access to it. As the term 'landscape' is used with multiple connotations, I commence this chapter by introducing contemporary views on landscape in human geography and related disciplines, and position this study's theoretical approach within the arena of modern-day landscape research. I also thematise the distinction between natural and cultural landscapes and challenge this analytical dichotomy (section 3.1). As the wider focus of this study is on people-place relationships, section 3.2 discusses the concepts of place and landscape and the tensions between them. Subsequently, section 3.3 focuses on a perspective on landscape that is of particular relevance in the context of the Godrevy case study: the understanding of landscapes as common goods and the question of who should have physical access to them. Section 3.4, introduces the strand of social-constructivist landscape research in detail, primarily following a theoretical approach by Kühne (2008, 2013). From this, I develop the analytical framework of this study. Section 3.5 at the end of this chapter summarises the theoretical approach with regard to its contributions for understanding local processes of climate change adaptation.

3.1 Contemporary Scientific Perspectives on Landscape

Both within and beyond the field of geography, the term 'landscape' has been used for centuries and with different, often changeable connotations. Landscape is therefore what Star & Griesemer (1989) would term a 'boundary object' associated with a multitude of meanings and used in a variety of disciplines and contexts (see also Ratter & Döring 2015). Although a discussion about the conceptual history of landscape in British and German human geography clearly exceeds the scope of this book, the "slipperiness of the […] concept" makes it necessary to position my understanding of landscapes amongst contemporary research tendencies to enable a fruitful use in relation to societal processes of climate adaptation. As it has particular relevance for the discussion of the empirical findings of this study, I also put emphasis on the ways in which the phenomenon of nature and the human-nature relationship are conceptualized from the different perspectives.

In geography, the phenomenon of landscape is viewed from both positivist and constructivist perspectives. Since the early 20th century, landscapes were predominantly seen as defined areas of land with objectively measurable characteristics and elements; a perspective still prevalent in strands of physical and human geography and related disciplines such as landscape ecology, environmental sciences or heritage management today. Gailing (2012a) argues that "[f]rom [such a] positivist point of view, a landscape exists independently – separate from the researcher, research methods and from an appraisal of social institutions" (ibid: 196). Many studies from physical geography thereby conceptualise the phenomenon of landscape as a synergy of the underlying geology, characteristics of its soil, assemblages of species of flora and fauna, ecosystems, and specific physical elements (cf. e.g. Fader et al. 2012, Wehberg et al. 2013). Landscape is thus understood from a natural scientific, positivist perspective as a definable spatial entity consisting of quantifiable material objects (Gailing 2012b, Antrop 2013, Kühne 2018). The focus of this perceptive does not, however, take into account the human dimensions of landscapes; the ways in which human actors perceive of and shape the landscapes they live and work in. The relevance of socio-cultural values and meanings attached to different elements of landscapes is widely overlooked, and the societal negotiation processes which constitute landscapes are inadequately considered (Greider & Garkovich 1994).

When shifting the focus on relationships between landscapes and people, contemporary human geography offers a variety of perspectives, although these cannot be presented here exhaustively. The following paragraphs will concentrate on those conceptualisations of landscape within Anglophone and German debates that are of relevance for the further course of this study. In a positivist tradition, when putting their human createdness at the centre of research, landscapes are understood as assemblages of objects formed through interactions between humans and their natural environment (Gailing 2012b, Antrop 2013, Kühne 2018). From this perspective, a landscape is seen as the "unique synthesis between the natural and cultural characteristics of a region" (Antrop 2013: 14), with research focusing on the genesis of its physical-material features as well as the societal processes resulting these features' development (ibid). Similar to natural scientific conceptualisations of the phenomenon, the materiality of landscapes is here taken as a given; in the center of research are the societal categorizations of different elements of such landscapes, the social values ascribed to them, and the analysis of conflicts in land use or landscape management (cf. section 1.2.).

Co-existing with these positivist perspectives on landscapes, a number of research strands move away from viewing landscapes as objectively existing entities and understand them rather as mental constructs on the basis of physical spaces. In the late 1980s, Cosgrove made an important contribution to this perspective by defining landscapes as "a way of looking at the world […] laden with particular cultural values, attitudes, ideologies and expectations" (Wylie 2011: 306, referring to Cosgrove & Daniels 1988). The author moved the phenomenon of landscape into the sphere of symbolism, connecting landscapes with topics of ideological loading as well as subjective meanings and agendas (ibid). Although considering social values and individual perceptions when conceptualizing landscape, the focus of this perspective was still predominantly on landscape as something visual which is interpreted on the basis of values. Picking up this notion of landscape as anchored in the subjective life-worlds of people, Ingold (1993) further develops the ideas of Cosgrove (1988) by adding a significant element he deemed was lacking: a focus on the everyday, lived-in notions of landscape. For him, what was most important about landscape was the fact that it is experienced through daily practices, through what the author terms 'dwelling in the landscape', and that the phenomenon of landscape and people's lives could not be separated from each other (Ingold 1993). The author therefore defines landscape in the following words:

> "Landscape, in short, is not a totality that you or anyone else can look *at*, is it rather the world *in which* we stand. […] For the landscape, to borrow a phrase from Merleau-Ponty, is not so much the object as 'the homeland of our thoughts'." (Ingold 2000: 207)

Much like Thompson et al. (2013) cited in this chapter's introductory quote, Ingold understands a landscape much rather as a concept, an idea, or a mental construct than as a distinct excerpt of the Earth's surface with definable boundaries (cf. Ingold 1993, 2000). Closely connecting landscapes to people's everyday lives and experiences, moreover, shifts the research focus towards "practices of landscaping – everyday things like walking, looking, gardening, driving building" (Wylie 2007: 11). In line with the works of Ingold (1993, 2000, 2010), Brace & Geoghegan (2010) understand landscape as a medium for visualising the effects of environmental change by grounding them in people's everyday life worlds (cf. section 3, introduction). Rose & Wylie (2006), on the other hand, underline the processuality of landscapes. They see landscapes in a "sense of space as a weaving and a relating, forever in the making" and highlight the changeability of landscapes (ibid: 476). This conceptualisation of landscape moves significantly beyond a visual understanding of the phenomenon and incorporates human practices, emotional and sensory experiences (Rose & Wiley 2006, Leyshon & Geoghegan 2012). "While the physical 'reality' of landscapes remains an important point of reference", these different perspectives concur in moving "human agency, symbolic representations, normative constructions of spatial images and – more generally – forms of cultural and social practice" into the spotlight of social scientific and human geographic landscape research (Gailing 2012a: 195). In this context 'construction' does not refer merely to the material (re-)shaping of different elements of the landscape, but also to the individual and collective processes of mental constitution through the attachment of values and symbols (Ingold 1993, Greider & Garkovich 1994, Gailing 2012a, Kühne 2018).

Landscapes in both German and Anglophone human geography today are understood as the product of complex braids of actors, policy decisions, and diverging interests in different facets of such landscapes (Soliva 2007, Kühne 2008, Brace & Geoghegan 2010, Gailing 2012b, Antrop 2013, Kühne 2018). Topics such as comparisons between different landscape perceptions (e.g. Soliva 2007), belonging and identity (e.g. Wylie 2009, Thompson et al. 2013), as well as the social practices constituting and reshaping landscapes (e.g. Brace & Geoghegan 2010) have become important subjects of investigation in this context. Contemporary studies consider the societal processes that underlie the (re-) shaping of landscapes, and the symbolism and attachments associated with certain landscapes or landscape features (Ingold 1993, Gailing 2012b, Antrop 2013, Kühne 2018).

The idea of landscapes as mental constructs rather than assemblages of physical objects is mirrored in the perspective which underlies this study: a constructivist approach to researching landscapes. Primarily developed by the

German geographers Kühne (2008, 2013) as well as Gailing (2012a, b) and Gailing & Kilper (2009), this perspective views landscapes as social constructions created in close relationship to physical spaces. The term 'construction' refers both to processes of physical-material construction, but more importantly to the ways in which different individuals and groups within society mentally construct landscapes (ibid). Thus, constructivist landscape research "conceptualises landscapes as lived and subjectively perceived constructs rather than focusing on the quantifiable, 'objective' characteristics of spaces" (Köpsel et al. 2016: 3). Which particular elements of a physical space are parts of a landscape and what meanings are attached to them, however, can vary considerably between different individuals and groups within a society. In this context, Gailing (2012a) importantly differentiates between the analytical constructions developed by scientists such as those presented above, and the individual and collective constructions held within a society but outside the realm of academia. Examples of such societal understandings of landscape are "environmental management policies and interventions [which] are designed and applied at the 'landscape scale'" such as in Natural England (2011) or classic Landscape Character Assessments (e.g. Cornwall Council 2011a) (Köpsel et al. 2016: 2), but also the prominent European Landscape Convention (Council of Europe 2000). Although focused on landscapes in an applied context, this convention demonstrates the operationalisability of a constructivist landscape concept in practice by defining landscapes as societal constructs *and* physical spaces (Council of Europe 2000, Kühne 2018). It is these societal constructions by local actors in Cornwall that stand in the focus of this study, approached through the scientific concept of landscape research and an analytical framework based on the works of Kühne (2008, 2018). To develop said analytical framework, I draw on Kühne's perspective on landscape and different dimensions of societal landscape construction in section 3.4. In the following, however, I build on the positivist and constructivist views on landscape presented above to challenge the distinction frequently made in human geography between so-called 'natural' and 'cultural' landscapes, and propose a perspective which transcends this conceptual divide.

3.1.1   'Natural' vs. 'Cultural' Landscapes from a Constructivist Perspective

In human geography, particularly in the German context, a differentiation has long been made between 'natural' and 'cultural' landscapes. Whereas natural landscapes are understood as stretches of land (largely) untouched by human activity, cultural landscapes are viewed as those modified through human practice; a standpoint grounded primarily in positivist conceptualizations of landscapes (Gailing & Kilper 2009, Kühne 2018). Important for thinking about natural or cultural landscapes and a critique of it is to note how 'natural' and 'cultural' are understood in this context, as these understandings contribute to the distinction (Gailing & Kilper 2009, Castree 2013, Kühne 2018). From a positivist perspective, nature is conceptualized as a given, as true and objective, and existing independently from its observer. Associated with flora, fauna, and the absence of human-made structures, natural landscapes are perceived of as dichotomous to cultural ones, to human activity and, above all, to urban spaces (Gailing 2014, Kühne 2014). However, speaking of a landscape in relation to nature thereby predominantly denotes not complete 'wilderness', but a stretch of land which is in one way or another managed through human interventions. Due to this focus on nature-like settings, human activities in natural landscapes are perceived as "disturbance to the ecological balance" and ecosystems, but also of an aesthetic ideal of what is considered as 'natural' (Antrop 2004, Gailing 2014).

On the other hand, cultural landscapes are understood from a positivist perspective as spatial units constructed materially through societal processes, putting emphasis on the fact that a certain stretch of land was shaped and modified by human activities (Schenk 2008, Gailing 2012b, Gailing 2014, Kühne 2018). Just as nature is an important ordering category in defining what is not human-made, the term 'cultural' underlines which kinds of landscapes are not natural ones. The use of the term 'cultural' thereby directs attention to the human efforts which lead to the genesis of the landscape in question (Schenk 2008). Hence, in a positivist tradition researching cultural landscapes means retrospectively identifying and comprehending the human activities which resulted in the development of certain landscape features or assemblages (Schenk 2008, Gailing 2014, Kühne 2018). Such human-made landscapes are often understood as being more organized and more ordered than natural ones, and planning activities in the context of cultural landscapes often have a strong focus on preserving a particular landscape character (Schenk 2008). Therefore, the wish for a harmonious relationship between nature and culture and the rejection of modernization of landscapes considered as important from a cultural heritage perspective (e.g. high rise or other modern buildings, large-scale traffic infrastructure, wind turbines) are often at the center of activities aimed at protecting cultural landscapes. Disciplines built on an understanding of landscapes in focus are monument preservation or historic geography (Gailing 2014). Gailing & Kilper (2009), on the other hand, have a more open and dynamic understanding of cultural landscapes which also pays attention to contemporary significance of cultural landscapes. For them, a "[c]ultural landscape can be understood as an endogenous potential which influences not only the living and working conditions of the people of a region, but also its economic competitiveness and attractiveness of tourism" (ibid: 114). Nonetheless, the distinction between natural and cultural landscapes remains a positivist endeavor and is based on the assumption that landscapes are objectively existing physical objects with certain features and characteristics.

However, from a constructivist perspective, this assumption does not hold as any meanings of nature are viewed as societally negotiated and as such are inherently "sociocultural phenomena" (Greider & Garkovich 1994: 5). In the introduction to his book 'Making Sense of Nature', Castree (2013: 6) argues that after all, "the very category of nature is part of the way we make sense of the world for ourselves". Likewise is a landscape is also such a category, and consequently any distinctions between natural and cultural landscapes are also part of how we make sense of the world (Gailing 2014). Moreover, as different subgroups within a society have their distinct constructs of nature that they take for granted and weave into their landscape perceptions, what nature 'is' always depends on the respective societal context (Greider & Garkovich 1994, Castree 2013). Therefore, "nature is one of our most genuine cultural creations" (Ruiz-Ballesteros et al. 2009: 147). From a constructivist perspective, the question is therefore not what natural and cultural landscapes are as objectively existing phenomena, but how nature is constructed in relation to humans and in different perspectives on landscape. With regard to the discussion of the empirical results of this study, it is therefore useful to examine some examples from contemporary literature of such constructions of nature.

Nature constructions in the Western world during the 19th and early 20th century were shaped by the idea of human domination of nature. Closely connected with industrialization and the excessive use of natural resources, the human system was constructed as separate from and superior to the natural. Nature was perceived as controllable, domesticable, and exploitable (Beck et al. 2001, Piechocki 2010, Gailing 2014). Perceptions of nature, however, changed in the 1960s and 1970s with the rise of the debate about global population growth and environmental pollution, resource scarcity and planetary limits (Meadows et al. 1974, Gebhard et al. 2007, Piechocki 2010, Castree 2013, Gailing 2014). On the one hand this debate gave rise to the idea that human life and activity depends on what nature does, thereby personifying nature and ascribing to it characteristics similar to free will or, at least, ways which are beyond the control of humans (Castree 2013). On the other hand, it resulted in the post-modernist perspective that nature and culture are hybrid, and our perceptions of both are societally constructed (Castree 2013, Kühne 2018). If "'nature' does not pre-exist as such, but rather is the result of a conceptual boundary being defined", then what is part of nature and what is not are societally negotiated ascriptions (Fall 2002: 243).

Following a social-constructivist understanding of landscape, landscapes are as a consequence never either 'natural' or 'cultural', but always constructed on the basis of specific understandings of the two. Therefore, positivist dichotomies such as 'landscape vs. city' or 'nature vs. culture' no longer hold (Kühne 2018). When viewing landscapes as individual and collective mental constructions on the basis of physical spaces, however, investigating the different perspectives on nature underlying these constructions provides significant insights into which elements of a physical space are considered part of a landscape on the on hand, and what implications result from them for landscape management on the other. The question of how nature is understood within a certain landscape construction therefore provides a useful perspective for analysing different narratives about the landscape with regards to the role of varying societal understandings of nature. Moreover, an analysis of how the human-nature relationship is perceived by different local actors is useful for understanding their viewpoints on landscape as a common good and the question of access for people to it (cf. section 3.3). This chapter will go on to address the relevance of viewing landscape as a commons in the context of this study. However, before doing so I will turn my attention to an important conceptual distinction that needs to be made when researching the role of landscape constructions for processes of climate change adaptation: the differentiation between the concepts of and place.

3.2 Landscape and Place – Concepts with Blurred Boundaries

Cresswell (2004) argues that "landscape and place are clearly highly interrelated terms and each definition is contested" (ibid: 12). Whereas in section 2.2 I introduced the usefulness of the concepts of place and place attachment for understanding processes of climate adaptation, I approach the empirical data gathered for this study primarily through the concept of landscape. Despite their distinct etymologies and intellectual histories the concepts of place and landscape are clearly interrelated. In human geography 'places' are understood as physical locations upon which a set of societal and individual values, beliefs, and attachments are superimposed (Tuan 1977, Adger et al. 2009, Devine-Wright 2014) (cf. section 2.2). Focused on people's individual perceptions of spaces, place is thus inherently anthropocentric and focused on human meaning-making (Wattchow 2013). The idea of landscape, however, has long been used across a variety of disciplines from geography such as planning, natural sciences, and ecology, with specific connotations and terminology featuring in each discipline. Human geographic landscape research in particular has classically concentrated primarily on the visual characteristics of rural areas as well as aesthetic preferences in different cultural contexts (Wattchow 2013, Antrop 2013, Cresswell 2015: 17) (see introduction to chapter 3). Landscapes have traditionally been understood as rather wide stretches of land, whilst places in current literature tend to be viewed as smaller geographical locations (Wattchow 2013). However, as outlined in section 3.1, the emergence of a constructivist perspective on landscape has shifted the focus to the individual and collective cognitive and material construction of landscapes. With an increasing focus on urban landscapes and the societal negotiation processes leading to their construction, current strands of landscape research turn from the discipline's traditional paths (Kühne 2018). Constructivist theorizations of landscape no longer differ widely from those of place. In recent years, therefore, the boundary between the concepts might best be regarded as blurred (see also Howard 2013: 3, Cresswell 2015: 18). Hunziker et al. (2007: 47) define landscapes as both a "physical 'space' for living but also a 'place' with its meanings and constitutions to societal identity". By closely linking landscapes with the notion of home and dwelling Wylie (2007), emphasizes the individual and visceral affective and embodied qualities in the concept of landscape, thus erasing the boundaries between the two theoretical stances. Agnew (2005), on the other hand, understands places both as locations in the sense of physical areas and as the sum of meanings, attachments, identities connected with them, a definition very similar to the constructivist definition of landscapes. An important aspect of what constitutes a place is always the assemblage of physical-material elements of where it is located and, reciprocally, landscapes are always also places in terms of people's attachments to and the experiences they associate with them. In addition, whereas landscapes can exist separately from a specific location in form of stereotypical ideas, places are always bound to the physical spaces that are loaded with meaning (Wattchow 2013). At the same time places, in contrast to landscapes, cannot be experienced from an 'outside' perspective. Perceiving a place is always inherently connected to 'insideness' (Wattchow 2013, see also Wylie 2007). Especially if the physical space serving as the basis for landscape construction is small in size, however, many of these distinctions do not hold. In this sense, the case study area of Godrevy in Cornwall can be understood in terms of both a place and a landscape.

Why then choose landscape as the main theoretical approach of this study, not place attachment? Firstly, the concept of landscape, as conceptualized in the sense of Kühne (2018) and Brace & Geoghegan (2010), sets a decidedly stronger focus on distinct perceptions of the materiality of the natural and built environments that underlie different societal approaches to climate change adaptation. Stemming from the field of environmental psychology, on the other hand, the idea of place attachment centers primarily on the social relations and emotional bonds that shape people's mental constructions of and attachment to certain places (Creswell 2004, Devine-Wright 2009). Although implicitly investigating people's emotional bonds with places, however, the analysis of the experiential and individual components of respective attachments, and the perceptions of the materiality of these places, in most cases remain superficial due to the predominantly quantitative methodology of such studies (see e.g. Devine-Wright 2009: 437, Scannell & Gifford 2013). In contrast to this, it was the central aim of my research in Cornwall to uncover local actors' perceptions of and personal and professional experiences with the local landscapes and, importantly, the physical-material effects of climate change to investigate different approaches to climate adaptation. In this context, the concept of landscape draws attention to the changing material spaces under investigation, thus clearly building a bridge between the abstract phenomenon of global climate change and people's subjective perceptions of the affected physical spaces. Especially the analytical approach by Kühne (2018) (cf. section 3.4) thereby provided me with much analytical depth in terms of subjective and lived experiences with the materiality of particular locations as well as notions of embodiment, dwelling and everyday practices. Such focus on the changes in the local actors' everyday surroundings and practices makes an intangible topic such as climate change graspable and brings it 'closer to home' (cf. introduction to chapter 3). Thus, the concept of landscape constitutes a rich theoretical tool for approaching this study's research objectives and for connecting local actors' perceptions of the material spaces they manage with distinct approaches to climate change adaptation. Additionally, the topic of landscape did not only serve as the theoretical approach of this study, but also as a valuable entry point to its case study area. Prominently featured in both policy-making and regional identity in the Cornwall, the issue of landscape formed a valuable link between the global phenomenon of climate change and the regional and local life worlds of those who participated in this study; a link much more specific to the geographical area under investigation than the more abstract concept of place attachment.

In summary, the "reciprocity between people and locations on the Earth's surface provides the reference point for all considerations of landscapes and places" (Wattchow 2013). Following Hunziker et al. (2007), Wylie (2007) and Howard (2013), I choose to not draw a clear line between landscape and place when analysing local actors' constructions of Cornwall and Godrevy. Whereas constructivist landscape research offers a helpful framework to theorize individual and collective constructions of landscapes, the concept of place emphasizes more the affective and personal bonds people have with where they live –taken up by Kühne (2013) in the emotional dimension of individual landscape constructions. Therefore, the individual and collective constructions identified from the empirical data of this study are constructions of both places and landscapes – the two concepts in their contemporary uses not being entirely distinguishable from each other.

Set against this background, the following section expands on a related concept that is particularly relevant for the empirical results of this study: the idea of landscape as a commons and the question of access to it.

3.3 Landscape as a Commons and the Question of Access

Especially when it comes to adaptation to an eroding coastline as is the case at Godrevy in west Cornwall (cf. section 1.3.3), an important question on the local level is who should have physical access to the landscape, and how to maintain this access (Köpsel & Walsh 2018). To address this question, a useful perspective is that of landscapes as common goods that a variety of publics have a right to access, walk through, and enjoy. The European Science Foundation & COST (2010) characterizes a landscape as "public space, accessible rural areas, nature reserves for the enjoyment of biodiversity, common land, seascapes and the coast, and sites of collective memory and identity" (ibid: 7). Underlying this definition is the idea that certain landscapes, or much rather those physical spaces which are the bases for landscape constructions, are common goods available for use by all members of a society. "Collectively used goods and services – whether roads, watercourses, lighting or […] landscapes – are a prerequisite of quality of life, economic productivity and a healthy environment of cities and regions" (Moss 2014: 458). While private goods are understood to be characterized by the exclusion of non-owners from using or consuming them, common goods are, with certain limitations, free for the public to use. A distinction is made in literature of those 'commons' into three kinds of publicly accessible goods:

Public goods: Goods "[e]veryone can derive benefits from […] without disturbing other uses" (Röhring & Gailing 2005, Gailing 2013: 17 referring to Ostrom 1990, Musgrave et al. 1994);

Club goods: Goods from which a certain groups of people is excluded, e.g. non-members from using a golf course (ibid);

Common pool resources: Those goods from which nobody can be excluded, such as a public lake or common land (ibid).

Davy (2009) further distinguishes between material common goods (such as public roads, car parks, or green spaces in cities) and immaterial ones (e.g. the image of a town or an exceptionally low crime rate). From a positivist perspective, the question of landscape as a common good is concerned with what formal institutions and rules (e.g. laws or ownership agreements) influence the property of the physical spaces that are considered landscapes; who has access and who does not; and who is responsible for managing such commons. Looking more closely at such ownership structures, however, many landscapes are not necessarily commons in the sense that their entire area is usable for everyone. If, however, "you draw your attention to the immaterial aspects of landscape, then you will be aware of landscapes as something collective and indivisible. Then landscape is a common" (Gailing 2013: 18). As examples for common goods in the context of landscape constructions, Gailing (2013) lists toponyms, agreed-upon constructions of where one landscape ends and another begins, traditions, spatial images and shared symbols or emotional attachments; all of which are commons in the sense that "any individual or collective actor can use the[se] immaterial factors" (ibid: 19). Additionally, many spaces constructed as landscapes can be seen as what the author terms 'regional political spaces' in the sense that the actors involved in managing these spaces have established "governance structures for the protection or the development" of respective landscapes (ibid: 19).

If the socio-economic, ecological and aesthetic functions of a physical space that are the basis for landscape constructions "are highly integrated with one another actors cannot reduce their activities to a single function without consequences for others" (Röhring & Gailing 2005: 2). Furthermore, if these actors have diverging interests, it is likely that conflicts arise (Gailing 2013). Although commons are important subjects of spatial planning, regional development and land-use policy, researchers in the field of human geography "have devoted surprisingly little attention" to their management in cases where a number of actors involved have dissenting interests (Young 2001: 284, Moss 2014). As outlined in 3.1.3, such diverging interests of different sub-societal groups in the context of landscape management are, for instance, nature protection, monument conservation, tourism, agriculture, and spatial planning (see Gailing 2012b). Moss (2014) highlights that in many local contexts, "actors are developing their own, place-specific solutions for organizing and cultivating commons", resulting in "multiple, shifting and contested geographies of the commons" (ibid: 458). To understand such local arrangements, it is vital to comprehend their local embeddedness, the constellation of the actors involved, as well as those actors' distinct standpoints in decision-making processes (Ostrom 1990, 2005, Moss 2014).

Röhring & Gailing (2005: 2) differentiate between socio-economic, ecological and aesthetic functions of common good landscapes. Physical-material landscape elements serving socio-economic functions are, for instance, fields for agricultural production, woodlands for forestry, or land for housing development. Such goods are in most cases privately owned and exclude non-owners from using them. Landscape elements such as flora and fauna or groundwater have ecological functions and are mostly common pool goods usable for anyone who is able to physically access them. Moreover, aesthetic functions such as beautiful scenery, are common goods enjoyable for everyone even from a distance (ibid).

The fact that "everyone can derive benefit from the high quality of [...] landscape without contributing to the preservation of this quality" thereby often constitutes particular challenges for those actors or organizations managing them (Röhring & Gailing 2005: 2); a circumstance described by Olson (1965) as the 'free-rider' problem: the possibility to enjoy a commons independent from whether the behaviour of the users is harmful to this common good or not. Closely connected to this problem is the question of who has, should have, or should not have access to a certain landscape; a question that is central in the context of this study's case of Godrevy, Cornwall (cf. section 1.3.3).

Mitchell (2008: 352) argues that "[a]ccess is about opportunities to be included within landscapes and, although being able physically to access these spaces is crucial to this, it is artificial to separate physical access from all the other forms of access to rural space that a healthy public presence requires, namely political and ideological claims". In line with Mitchell's viewpoint, when speaking about access in the present context I do not focus on the discussion about 'accessibility' in terms of the possibility to physically access particular sections of a landscape[9]. It is rather the normative question of whether and in which form people *should* or *should not* have access to a certain landscape per se that is of importance here. From a social-constructivist perspective, the question of who should have access to a certain physical space and who should not is a clearly normative one, and its answer is rooted in these actors' individual or sub-societal constructions of the respective landscape. As it also has important implications for the focus of landscape management, it is worth taking a closer look at existing literature about the question of 'letting people in' versus 'keeping people out' when it comes to the construction of landscapes.

Whereas in pre-modern times human communities had to protect themselves from nature (in terms of wilderness, the unknown, and dangerous predators), industrialization and urbanization in the 19th century reversed this relationship. Particularly in the United Kingdom, access to the countryside has a long-standing tradition: the rapid industrialization of the early 20th century gave rise to "growing appreciation of the great outdoors, the benefits of physical exercise, and the feeling of freedom and of spiritual renewal gained from open-air recreation", fostering the demand for a right of access to both publicly and privately owned rural areas (UK National Parks 2017). Nowadays, many areas which are considered 'natural' are perceived as in need of protection from humans (Piechocki 2010).

Contemporary counter-positions to the exclusive agendas of nature protection sites are today's landscapes of tourism and agriculture, representing sectoral landscape constructions contrary to those focused on keeping people out.

---

9   For a review of the academic discussion around accessibility see SEMM & PALANG (2010)

Viewing landscapes as bases for the maximization of profit, the sectors of both tourism and agriculture thereby rely on including humans as parts of the landscape. They therefore centre their landscape management activities on enabling people to access and make active use of the physical spaces underlying those landscapes (Gailing 2012b). Especially in areas with a strong tourism sector those physical spaces (in the sense of these area's natural capital) play a vital role for the local economy and their image as a tourism destination (Kühne 2018: 248ff). Accommodating visitors in the landscape while not impairing its visual attractiveness is one of the central management challenges, especially where the landscape construction by the tourism industry competes with those of nature protection or monument conservation (Gailing 2012b, Kühne 2018). Agriculture, on the other hand, makes use of physical spaces by ploughing soils, cultivating food or grazing cattle; activities which clearly require active human interventions in the landscape (see e.g. Geoghegan & Leyshon 2012). The question of access is therefore important not only for different actors involved in managing landscapes, but also for local populations affected by different approaches to said management (Sandell 2005). However, despite the important role of access in different approaches to landscape management and in potential conflicts between actors, studies addressing this issue in the context of social scientific or human geographic research are scarce. Furthermore, Moss (2014) criticizes the neglect of "the economic, social, or cultural geographies which also shape the production, use and regulation of commons" regarding and exceeding the question of access (ibid: 460). When considering landscapes as commons, I argue in line with Moss that, in the context of climate change adaptation, it is therefore important to comprehend the material and mental constructedness of landscapes. This includes considering what actors construct landscapes as commons and what actors do not; what different perspectives on the access question can be found among actors involved in processes of physically-materially (re-)shaping landscapes; and whether and how contrasting viewpoints lead to conflict among local actors with differing adaptation approaches (cf. Köpsel & Walsh 2018).

To develop a framework for the analysis of this study's findings from Cornwall, the following section introduces in detail this study's perspective on landscape. To this end, I highlight the different individual and collective dimensions of landscape construction (3.4.1), give an overview of such constructions that can be found in existing literature (3.4.2), and expand on the process of the societal negotiation of landscapes (3.4.3).

3.4 Constructivist Landscape Research

In the summary of his 2013 book about landscape and landscape practice, Kühne proposes that the 'landscape perspective' should be put at the center of all considerations regarding the societal handling of space; after all, he argues, the concept of landscape is anchored in the life-world of people, whereas space constitutes a very abstract way of thinking (Kühne 2018: 325). Similarly, the European Science Foundation & COST (2010) underline that landscape research is suitable to "holistically address major issues in the social and physical transformation of land, space and environment, and in past, present and future, relevant for addressing these challenges" (ibid: 1).

Greider & Garkovich (1994) define landscapes as "symbolic environments created by human acts of conferring meaning to nature and the environment, of giving the environment definition and form from a particular angle of vision and through a specific filter of values and beliefs" (ibid: 1). Thus, "reality is socially defined and the meanings of things within this reality are created by people" (Greider & Garkovich 1994: 9). Following this perspective, different ideas about what landscapes are and what elements compose them are learned and shared by members of societies through processes of socialization. Such beliefs can be viewed as the stock of a society's 'general knowledge': taken-for-granted assumptions about phenomena such as landscape which in most everyday situations are "so implicitly obvious to the individual, that it is indistinguishable from the person's self-definition" (Greider & Garkovich 1994: 2). Hence, social constructivism "is underpinned by the notion that no final, stable 'truths' exist independent of the actors who construct them" (Leyshon & Walker forthcoming: 8). At the same time this means that different groups within a society construct phenomena differently according to their subjective truths and socializations (Greider & Garkovich 1994, Leyshon & Walker forthcoming). Gailing (2012a) argues that from a constructivist perspective, landscapes are "first of all dependent on subjective prerequisites and [are] deeply influenced by cultural factors such as norms, values, ideologies, or attributions of meaning" (ibid: 197). Such cultural factors do not only define who we are and how we see ourselves, but also the relationships we have with the place we live in: what we hope for it to become and how we choose to shape it (Greider & Garkovich 1994). Therefore, Wylie (2007) underlines that landscapes are to be understood as both 'real' physical stretches of land that one can travel through and find oneself in, but also people's perceptions of those spaces from their own perspectives (ibid: 7). Landscape thus originates at the interface between physical objects, individuals, and society (Kühne 2018: 55).

From a constructivist perspective, the construction of a landscape "is a social process, but it is also the product of long-term natural and human processes in the past; it is subject to continuous change" (Ingold 1993, Hunziker et al. 2007: 5). What is understood as landscape and by whom, what elements of the natural and built environment play a role in it, and in what constellation they form a landscape are the result of societal negotiations (Kühne 2018). Based on the fact that cultural values, norms, and attachments differ both between societies and between groups and individuals within societies (O'Brien & Wolf 2010, Adger et al. 2013), the term 'landscape' is loaded with a multitude of meanings and interpretations among different actors (Greider & Garkovich 1994: 2). It is the aim of such landscape research to investigate and explain "what people mean when they say 'landscape'", and what collective and individual values underlie their perspectives (Kühne 2018: 26). Its aim is thus not to determine what a landscape is and what it consists of, but to understand how different people and groups within a society make sense of the phenomenon of landscape from their perspectives and based on their knowledge, cultural values, and personal experiences (Gailing & Leibenath 2015, Revill 2016, Kühne 2018). On such conceptual level, 'landscape' "is a vital, synergistic concept which opens up ways of thinking about many of the problems which beset our contemporary world, such as climate change, social alienation, environmental degradation, loss of biodiversity and destruction of heritage" (Howard et al. 2013: foreword) (see also Greider & Garkovich 1994, Geoghegan & Leyshon 2012, Köpsel et al. 2016).

A particular strand of landscape research was developed in Germany in the past decade primarily by O. Kühne (2008, 2009, 2018) and, working with similar ideas, L. Gailing (2012, 2013) and H. Kilper (2013): social-constructivist landscape research. Here, landscapes are understood as an interplay between a physical space and the ideas that are societally constructed upon this physical space (Greider & Garkovich 1994, Kühne 2008, Howard 2013) (cf. section 3.1). The focus of this strand of landscape research lies on investigating what elements of a physical space different groups within a society perceive of as a landscape, what values and meanings they attach to them and, importantly, what does not belong to respective landscape for them (Kühne 2018). Such groups within a society can, for instance, be lay people, scientists, planners or politicians (Leibenath 2013). In the following subchapter, I present how Kühne (2008, 2018) and colleagues conceptualize the societal process of landscape construction with its different components, and propose an analytical framework through which to view the empirical results of this study.

3.4.1 Dimensions of Landscape Construction – Towards an Analytical Framework

As sketched above, it is argued here that constructs of landscapes are superimposed upon physical spaces by societal through the ascription of meaning, symbols, cultural values, and emotional attachments (Ingold 1993, Gailing 2012b, Kühne 2018). At first glance, the social-constructivist understanding of landscapes is thus not entirely different from the concept of place; however, the two schools of thought underlying place and landscape research indeed are (cf. section 3.2). However, as this perspective ascribes an important role to the physical environment which serves as the basis for these constructions, it contrasts with a solely discursive approach in which material aspects only become socially relevant if they are communicated through discourse (comp. Gailing & Leibenath 2015). O. Kühne (2008, 2018) proposes four dimensions of societal landscape construction which, together with their sub-categories and the considerations of other authors from this field, will make up the analytical framework of this study (see Figure 6). Against the background of this perspective on landscape, I now expand on the different dimensions of landscape construction proposed by Kühne (2008, 2018) to subsequently present a framework for the analysis of the empirical data of this study anchored in social-constructivist landscape research.

In his 2008 book "Distinktion–Macht–Landschaft" (Engl.: ‚Distinction–Power–Landscape‘), Kühne argues that societal constructions of landscapes consist of four components (see Figure 5, p. 55):

1. **The Societal Landscape** = the shared societal[10] knowledge about what a landscape is and what elements it consists of;
2. **The Individually Updated Landscape** = individual interpretations of the societal landscape;
3. **The External Physical Space** = the physical-material basis for landscape constructions;
4. **The Appropriated Physical Space** = those elements of a physical space that are considered in a landscape construction.

Based on these four dimensions, the construction of landscapes is understood as both a societal and individual processes which is undertaken on the basis of a physical space, certain elements of which are considered as part of such landscape (Kühne 2018). Beginning with the 'societal landscape', I will now outline each of these dimensions to bring them together in a framework for analysis.

---

10   Using the Oxford Dictionary definition of 'society', I understand the term as referring to a "community of people living in a particular country or region and having [*a certain common basis of*] shared customs, laws, and organizations" (OXFORD DICTIONARY ONLINE 2017b). Using 'society' as an umbrella term, I however acknowledge that no group of people is ever homogenous, and customs and values might vary widely between different groups within such society (comp. O'BRIEN 2009, ADGER ET AL 2013)

1. The Societal Landscape

What Kühne (2008, 2018) terms the 'societal landscape' is the agreed-upon knowledge within a society about what is perceived to be a landscape, what elements it consists of, and what cultural values and normative associations are connected with it. This collective knowledge refers to the arrangements of elements of a physical space that are considered a landscape, the ascription of symbols and values to certain objects in this landscape, and the question what is of aesthetic value and what is not (Bruns & Kühne 2013). This knowledge is learned through socialization within society, for example at school, from parents and relatives and within peer groups, and is shared among members of this society through media such as television, but also through communication about landscapes. Shaping both the societal perception of what a landscape is and expectations for its future, this knowledge is inherently normative (Kühne 2018)[11]. Closely related to this common stock of knowledge are what Kühne (2018: 57) terms 'societal emotional conventions' (German: gesellschaftliche Gefühlskonventionen), which dictate that certain feelings are to be associated with landscapes, for instance a sense of freedom connected to the vast expense of a coastal landscape. Such shared landscape constructions are the "[r]egional basis of individual and collective identities, of regional utopia […], of regional ideologies, or of the material and immaterial heritage of a bounded space" (Gailing 2012a: 198).

The societal landscape is described by Kühne (2018) as the totality of landscape constructions within a society with all their interpretative patters, emotional associations, and constitutive elements. As hardly any society is homogenous, different groups and individuals "have constructed different symbolic meanings for the land thereby creating different landscapes that reflect their definition of themselves" (Greider & Garkovich 1994: 12). Such 'sub-societal' (German: teilgesellschaftlich, Kühne 2018: 58, 64) groups can be different sectors within society such as nature protection or tourism (see e.g. Gailing 2012a), but also different ethnic groups or people with varying durations of residence in an area (Kühne 2018: 239). This shared landscape knowledge and the different individual landscape constructions held by members of a society are reciprocally interdependent. Only the ability to draw on a shared stock of ideas about what landscapes are and what values are associated with them enables individuals to develop their own constructions by building on those shared ones (ibid). In the course of this study, the analysis of these collective constructions gives insight

---

11   'Normative' is here not understood in the sense of, for example, JACKSON (2005 [1984]) who claims that there is one ideal type of landscape which people are striving for. Rather, 'normative' is taken to refer to the fact that different collective and individual constructions of landscapes result from different normative stances on how these landscapes should be managed and what elements of these landscapes should be protected.

into those perceptions of the Cornish landscapes that are agreed-upon common denominators among local actors in Cornwall and at Godrevy, and which are also closely connected with Cornish regional identity.

2. The Individually Updated Landscape

Kühne (2013: 64) defines the 'individually updated landscape' as the personal interpretative patterns of individual members of a society, including their symbols, emotional attachments, and notions of identity and belonging. These individual landscape constructions are influenced decisively by the societal landscape in which one was socialized, but also by personal experiences, preferences, as well as educational background (Bruns & Kühne 2013, Kühne 2018). Drawing among others on Turner (1996), Wagner (1997) and Mitchell (2002), Kühne (2018: 59) identifies five dimensions which make up such individually updated landscape constructions:

a. The **symbolic dimension** refers to the symbolic meanings attached to entire landscapes or smaller elements of it.
b. The **aesthetic dimension** primarily draws on the scheme of 'pretty/ugly' based on socially learned patterns of interpretation. In Western Europe, many aesthetic landscape preferences today are still influenced by Romanticist painters who laid the foundations for what we perceive as picturesque and beautiful landscapes (Wattchow 2013).
c. The **cognitive dimension** consists of individual knowledge about a space constructed as a landscape and can either be learned through socialization (e.g. at school) or in the course of professional training (e.g. knowledge from soil science or heritage studies).
d. With regard to the **emotional dimension**, Kühne (2018) understands landscapes as 'projection screens' for societal emotional conventions as well as emotional attachments based on personal experiences tied to certain landscape elements. Therewith, individual landscape constructions "reflect deeply who we are and [are] a storehouse of private and collective memories" (Taylor 2008: 3).
e. Lastly, the **normative dimension** of landscape construction refers to the current and desired state of a landscape based on societal patterns of interpretation and meaning (Kühne 2018: 59).

Considering these five dimensions, individual landscape constructions with their different components are highly based on a person's context of "whatever tasks they are currently engaged in and expectations of the future as well as the past" (Ward Thompson 2013: 31). In the course of this study, the analysis of the individually updated landscape constructions brings to light how the interviewees in Cornwall personally perceive the landscapes they live and work in, and constitute a useful ordering scheme also for the analysis of the sub-societal narratives.

3. The External Physical Space

The external physical space is viewed by Kühne (2018) as the basis for landscape constructions and is located outside of the individual. From the social-constructivist perspective which is focused on both the physical and mental facets of landscapes, the external physical space constitutes the material dimension of landscapes. Although its elements might be perceived and interpreted differently by different individuals, this physical space exists largely independent from the viewer (Gailing 2012a, Kilper & Gailing 2013, Kühne 2018). The external space thus refers to the spatially-relational arrangement of objects in an area per se, no matter which of those objects are taken into consideration in the construction of landscapes (Bruns & Kühne 2013). In this sense, "nature and the environment [...] become landscapes through societal interaction and negotiation" (Greider & Garkovich 1994: 19). Landscape change is then seen as a coming together of changing mental constructions of landscapes and the change of the physical-material spaces which underlie those constructions, either through natural developments or human activity (Kilper & Gailing 2013). While clearly acknowledging their materiality, the focus of this strand of landscape research nonetheless lies on the societal and individual constructions created *upon* those physical spaces (Kühne 2008, Gailing 2012a, Leibenath 2013, Kilper & Gailing 2013, Kühne 2018). Often neglected in more discourse-oriented approaches to understanding processes of landscape construction (e.g. Leibenath 2014), from the social-constructivist angle "all other dimensions of the construction of landscape refer to the materiality of physical objects in a spatial context" (Gailing 2012a: 197). Such physical objects can be structures such as houses, cars, or vegetation, but also current as well as historic land use patterns. Gailing (2012: 197) views this external, physical dimension of landscapes as "a unique result of the intended and unintended interaction of human beings with their physical surroundings".

4. The Appropriated Physical Space

The appropriated physical space is the assemblage of material objects in a physical space which are 'used' to construct societal and individually updated landscapes. It is this dimension which is colloquially referred to with the term 'landscape'. 'Appropriation' thereby refers to the attachment of meanings to certain elements of a physical space and to their assemblage based on learned societal interpretative patterns and shared knowledge about landscape (Kühne 2018: 64).

The appropriation of the physical space is a selective process: not all material objects in a certain area are considered as part of a landscape, and what objects are varies between different groups and individuals within a society. Therefore, "any physical place has the potential to embody multiple landscapes, each of which is grounded in the [self-]definition of those who encounter that place" (Greider & Garkovich 1994: 2, Kühne 2018). Based on the three dimensions outlined above, Kühne (2018) distinguishes between three types of appropriated landscapes:

I.   **The societally appropriated landscape** = those objects of a physical space which all members of a society agree upon as part of a landscape;

II.   **The sub-societally appropriated landscape** = those objects of a physical space considered as a landscape due to belonging to a certain group or social milieu, residence in a certain region, or professional education from a certain sector of society (e.g. tourism or nature protection);

III. **The individually appropriated landscape** = those objects which are being used to construct a landscape by individuals based on societal landscape knowledge as well as personal education and experiences.

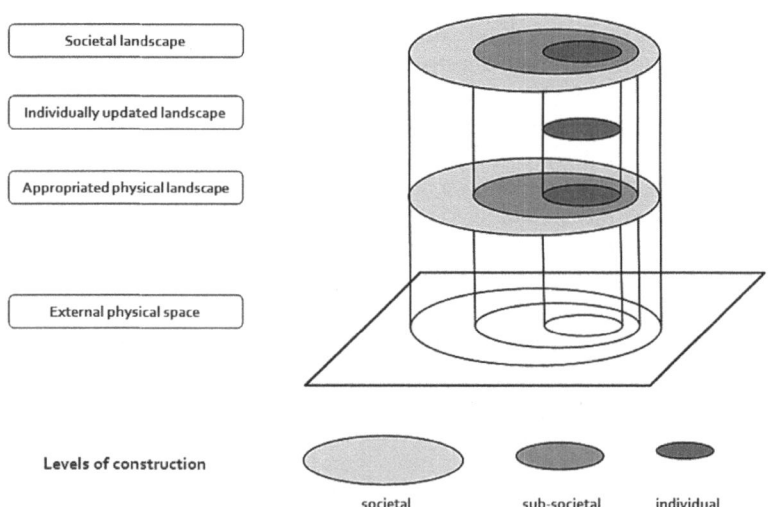

Figure 5: Dimensions of Societal Landscape Construction
Source: Kühne (2018: 66), modified

As the material objects of physical spaces are often rearranged through processes of planning, building, or re-purposing of land, such appropriated landscapes are unstable constructs and in a constant process of adaptation to new conditions. Moreover, the societal interpretive patterns for these physical objects are also prone to continuous change (see e.g. Gailing 2012a on landscape and informal institutions, Kühne 2018).

A further facet of appropriated landscapes is addressed by Wylie (2007) and Howard (2013): the question of 'insideness' and 'outsideness' of individuals with respect to their landscape constructions. The authors describe as 'insdeness/out-sideness' two different kinds of relationship of the constructing individual with the landscape. 'Insiders', Wylie (2007) states, accumulate knowledge about a landscape by immersing themselves in the physical space it is based on and "live[ ] it, touch[ ] it, explore[ ] it on foot – inhabit[ ] it" (ibid: 5). Wylie (2007, 2009) connects this insideness with what Ingold (1993, 2010) terms 'dwelling'; the experience of a landscape through "being-in-the-world", a "co-presence of self and landscape", and "bodily engagements, encounters and inhabitations" of the physical elements of a landscape (Wylie 2009: 278) (cf. section 3.1). Also closely connected with notions of emotional attachment, the inside perspective on landscapes shows similarities with the concepts of place and place attachment (cf. section 3.2). 'Outsiders' to the landscape, on the other hand, take on the role of observers by standing back and perceiving it from a distance, without relation to themselves. Instead of a bodily experience, landscape is perceived as an external, observable entity (ibid: 5). Such an outside perspective is often accompanied by the construction of landscapes based on their aesthetic charac-teristics as well as notions of romanticism and nostalgia (Howard 2013).

Important to note is that the four dimensions of construction by Kühne (2008, 2013) are neither used consciously or purposefully by those who construct landscapes, nor are they to be understood as chronological steps in a linear con-struction process. In contrast, they are an organizing scheme to analyse landscape construction from a social scientific perspective and tools in the sense-making process of the researcher. Landscapes in the context of this study are therefore understood from an analytical perspective as individual and collective (societal and sub-societal) mental constructs *on the basis of* a physical-material space. With regards to the empirical data of this study, this perspective on landscape means inquiring about

- how landscape is constructed **societally** in Cornwall and Godrevy (in the following: 'collectively shared landscape constructions');
- what **societal sub-groups** construct this landscapes in what way (in the fol-lowing: 'sub-societal landscape constructions');

- what **individually updated** landscape constructions (in the following: 'individual landscape constructions') can be found among local actors in landscape management;
- and in what **imperatives for landscape management** and climate change adaptation result from these different constructions.

A Framework for Analysis

Figure 6 illustrates the conceptual framework that has been developed to constitute the basis for the analysis of this study's empirical data. It brings together the four components of societal landscape construction by Kühne (2013) (see points 1-4, p. 50/51), the five dimensions of individual construction (see points a-e, p. 52/53), and their imperatives for landscape management and climate adaptation.

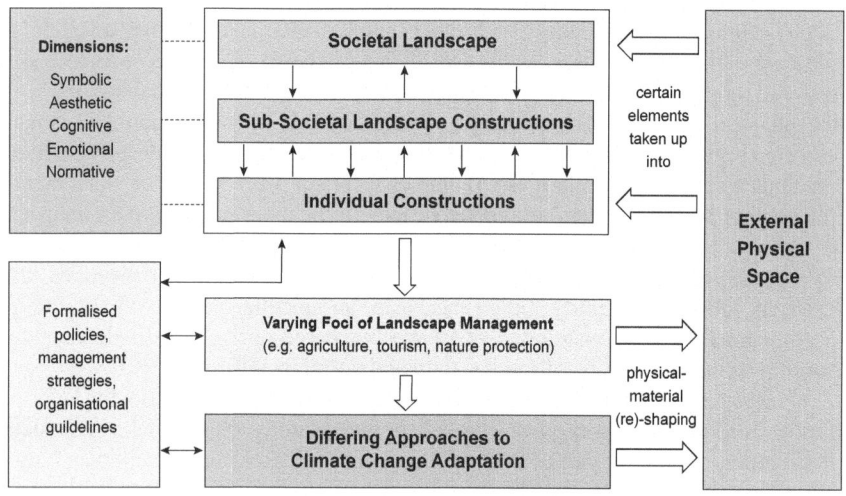

Figure 6: Analytical Framework
Source: own figure, based on Kühne 2008, 2018; Gailing 2012a, 2012b

The different components of landscape construction based on Kühne (2008, 2013) form the center of the framework: societal, sub-societal as well as individual landscape construction which stand in reciprocal relationships with each other, and each with their symbolic, aesthetic, cognitive, emotional, and normative dimensions. These constructions are based on those elements of the physical space which are considered as part of a landscape (Kilper & Gailing 2013, Kühne 2018). Varying sub-societal foci of landscape management result from the interplay between the individual and collective levels of landscape construction. Partially formalised in terms of organizational guidelines and strategy documents by the actors involved, these management foci can constitute opposing viewpoints in negotiation processes around land use or management of the physical-material features of the landscape (Greider & Garkovich 1994, Gailing 2012b, Kühne 2018). As this study argues such varying foci of landscape management as well as such formalised organizational policies and strategies result in differing approaches in how to materially adapt a certain landscape and its external physical space to a changing climate.

3.4.2 Landscape Constructions in Contemporary Literature

To later contextualize the empirical findings of this study, Table 1 summarizes different societal landscape constructions found in present-day literature with reference to Micheel (2013) and Kühne (2018), but also with reference to their colleagues who researched constructions of landscapes both in the German as well as in other contexts[12].

In addition to these broader categories, other authors make further distinctions when it comes to the construction of landscapes by certain groups of society. Gailing (2012b), for instance, analyses how different sectoral systems such as tourism, nature protection, agriculture or spatial planning conceptualize landscape, and what informal institutions shape these constructions. Each of these sectors has their own formalized guidelines and policy frameworks for landscape management, focusing on varying physical and cognitive elements of such landscapes. Whereas the sector of nature protection has been shown to construct landscapes based on ecological considerations and with a strong notion of traditional aesthetic values, the tourism sector has been shown to view landscapes much rather as specific destinations with a unique local image and marketing potential (ibid) (see Table 2).

The Landscapes of Lay People and Experts

Another distinction is made by Howard (2013) and Kühne (2018) who differentiate between the landscape constructions of experts and lay people. However, lay and expert knowledge are not understood as distinctly separated bodies of knowledge; they much rather represent different sections of the same stock of societal knowledge (cf. section 2.4). The landscape constructions of lay people are thereby closely connected with what Wylie (2007) describes as the notion of 'dwelling' – the bodily experience of landscapes through being-in-the-world, of being in the physical spaces which underlie those landscape constructions (cf.

---

12   I am referencing here primarily those publications which give an overview over multiple landscape constructions based on empirical studies. For detailed references of the individual studies underlying these compilations, please see respective sources.

Table 1:   Landscape Constructions in Contemporary Literature

Construction	Constituting elements	Reference(s)
Landscape as natural	Natural environment; local flora and fauna; green; vegetation and relief, geology; pristine	Micheel (2013) Hokema (2012), Gailing (2012), Kühne (2018)
Landscape as rural	Landscape as *not* urban/infrastructure/industry/settlements; dichotomy landscape ↔ city; suburbanization as central topic	Hoskins (1955), Micheel (2013), Hokema (2012), Kühne (2018)
Landscape as home	Positive connotations with landscape; notions of attachment and romanticism; landscape as self-definition of the individual	Ratter & Gee (2012), Micheel (2013), Kühne (2006, 2018)
Landscape as beautiful and good	Aesthetic notions of landscape; poetry and painting; engineered structures as disturbances; peace and quiet	Hook (2008), Micheel (2013), Kühne (2018)
Landscape as recreation	Landscape as peaceful and quiet, therapeutic landscapes; place for outdoor activities	Micheel (2013)
Landscape as visuality	That which can be seen from a certain location; also other sensory experiences as part of landscape; 'outside' perspective (see below)	Wylie (2007), Howard (2013), Kühne (2006, 2018)
Landscape as memory	Childhood memories, emotional attachments with certain landscape elements	Micheel (2013), Ratter & Gee (2012)
Landscape as stereotypical	Landscape as stereotypical assemblage of features (woodlands, rivers, moors, paths, etc.); e.g. 'the mountain landscape'	Homeka (2012), Kühne (2018)
Landscape as an object	Landscape as an external, material reality; landscape perceived as re-shapeable and plannable object	Micheel (2013), Kühne (2018)
Landscape as region	Landscape as an area with distinct location names and limits; container space perspective	Micheel (2013), Gailing (2012), Kühne (2018)

Source: own compilation

Table 2: Sectoral landscape constructions by Gailing (2012b)

	Nature protection	Monument conservation	Tourism	Agricultural development	Spatial planning
Landscape constructed as	Ecological assemblages; traditional aesthetic values; local distinctiveness	Historic cultural landscape; monumental; iconic land marks	Destination with specific image; subject to marketing and economy	Basis for agricultural interventions and rural development	Potential for development; justification for strategies; space of land use conflicts
Basic values	Ecological and cultural value; economic tendencies	Cultural-historic values; preservation	Economic values; foundation for value added	Economic values; foundation for value added	Diffuse set of values

Source: Gailing (2012b), modified

section 3.4.1). In this context, Vanderheyden et al. (2014) underline the important role which these ordinary, day-to-day landscape constructions play in people's everyday life worlds. Such 'everyday landscapes' (ibid: 592) are those familiar-landscapes that people live and work in, and to which they are attached emotionally in the sense of a feeling of belonging. Which values and meanings are associated with certain elements of these landscapes is learned through socialization during childhood and adolescence and through experiences with these landscapes as places of residence (Kühne 2018). The aesthetic qualities of such landscapes thereby move into the background behind memories of local language, noises, scents, colours, gestures, and feelings. It is due to these memories and emotional attachments that local residents often reject the alteration of the physical basis of 'their' landscape and are willing to pay high costs to conserve what is familiar and dear to them (Hüppauf 2007, Kühne 2018). Kühne (2018) contrasts these local 'everyday landscapes' in the perception of lay people with what he terms 'stereotypical landscapes': those lay landscape constructions that are influenced by art, painting and marketing and which, in many cases, a person has never physically experienced. Aesthetic values stand in the center of these landscapes; connected to, for example, idealistic ideas of mountain or coastal landscapes in locations far from home (Kühne 2018).

Howard (2013) and Kühne (2018) define the 'landscapes of experts' as those landscape constructions that are learned at adult age in the form of specific courses at university or professional training (e.g. geography, landscape planning, architecture, conservation, urban planning). The specific knowledge informing 'expert' landscape constructions is gathered systematically and based on certain pre-defined interpretative patterns of sub-groups of society. Each of these sub-societal groups has their own terms of reference, discourses, definitions and, in many cases, specific conferences and journals to distribute the respective knowledge (ibid: 283). These expert constructions are often institutionalized through university degrees, memberships in certain associations, or occupation by certain organizations such as national parks or planning authorities. In terms of 'knowledge is power', this additional knowledge about certain elements of a landscape and their interconnections tends to make 'experts' more powerful in their professional realm and, at the same time, often degrades forms of knowledge not recognized within a certain field. Little significance is paid in most expert concepts of landscapes to aspects of emotions and attachments, in turn devaluing experiential or local knowledge about landscapes (Howard 2013) (cf. section 2.3).

Academic theories about such landscape constructions thereby constitute a specific form of expert knowledge and are organizing schemes applied by researchers to make sense of the various forms in which both 'experts' and lay people construct landscapes (Kühne 2018). Although consisting of particular, professional knowledge, however, the landscapes constructions of 'experts' nonetheless are decisively based on the 'lay' constructions learned during childhood and youth. How 'experts', in this case professionals working in different fields of landscape and environmental management, construct certain landscapes hence depends on both the specific knowledge learned through training and education, and their personal, intimate experiences with and perceptions of this landscape (Howard 2013, Köpsel et al. 2016, Kühne 2018). An example of such expert constructions is what Leyshon & Walker (forthcoming) term framings: distinct understandings of the landscape by different actors in environmental management regarding varying approaches to landscape and risk management (cf. section 4.1.2). Thus, the different kinds of landscape construction introduced so far, scientific and lay ones as well as those from marketing and art, are in a reciprocal relationship with each other, influencing people's landscape perceptions and being influenced by them in turn (Gailing 2012a).

If different landscape constructions – such as those of experts and lay people, but also those of other sub-societal groups – confront each other in times of landscape change, Greider & Garkovich (1994) argue, a societal process begins in which the meanings and values attached to those landscapes are (re-)

negotiated between the different actors involved. The following section will elaborate upon such processes of negotiating landscapes.

3.4.3 Negotiating Landscapes

> *"Every river is more than just one river.*
> *Every rock is more than just one rock."*
> *(Greider & Garkovitch 1994: 1)*

Any physical space that is predominantly used for one purpose is, in most cases, also used for another (e.g. agrarian use, species protection, recreation, etc.). Likewise, any physical space upon which landscapes are societally constructed "has multiple meanings, [*and*] these meanings are symbolic reflections of how people define themselves" (ibid: 14). In a similar context, Cloke & Little (2005) speak of "the inevitability that different representations of culturally constructed countrysides will be mutually contesting in one way or the other" (ibid: 1). Co-existing societal constructions of the same physical space can easily lead to conflict among different groups within a society, especially when this space is undergoing change (Greider & Garkovich 1994, Cresswell 2015, Kühne 2018).

Greider & Garkovitch (1994: 14) find that when physical-material change occurs or is about to occur in a physical space, every group and individual within a society interprets this change on the grounds of their self-definition (closely linked to their construction of this landscape), and enters into a process of re-negotiating this self-definition in the face of the change about to happen (cf. introductory quote above). In this light, "[w]hat scientists or developers may define as a simple modification of the natural environment [...] may be defined as a threat to the fundamental meaning of [another] group's lifeworld". Such change seldom does *not* lead to disagreement or conflict between sub-societal groups with differing constructions of a landscape. Therefore, in most cases social impacts "occur prior to the actual implementation of the development project of environmental change" (Greider & Garkovich 1994: 14). When faced with change, a process of negotiation begins within and between the different societal groups involved. These negotiations take place within the context of these groups' existing landscape constructions, guiding their ambitions for the outcomes of the change.

On the individual level, the re-shaping of landscapes is hence always grounded in what one already knows about the landscape in question, and how one learned in a certain societal context to perceive and evaluate landscapes, making the perception of landscape change part of a person's individually updates landscape construction (Kühne 2018). Landscape construction, in both mental and physical-material senses, mostly does not happen in order to produce aesthetically appealing stretches of land, but is interwoven with a variety of interests of sub-societal groups following their own logics (ibid). Gailing & Röhring (2008) and Gailing (2012b) list as such sub-societal groups sectors like agropolitics, the water economy, tourism, or nature protection, all with varying interests in landscapes. The intended outcomes of landscape management processes of such groups are thereby driven by their primary interest in the physical space that is the basis for their landscape constructions. Such desired outcomes can, for example, be economic interests (e.g. real estate development or agricultural production), societal uses (e.g. recreation) or political agendas (e.g. development of renewable energy infrastructure or nature protection). Farmers, for instance, will prefer landscapes that are suited for agricultural production (Kühne 2018). The economic resources, political standpoints, as well as positions of power of the actors involved thereby decisively influence the negotiation process (cf. Greider & Garkovitch 1994). The main conflict potential is identified by Kühne (2018: 247) to be what he understands as 'interest-driven landscape constructions' with internal logics and implications for landscape management (see also Hunziker 2010). He thereby differentiates between four main interests in landscapes often represented in negotiations about land use and development:

- **Tradition:** Focus on the conservation of traditional landscape scenery, oriented towards an idealized idea of a historic landscape.
- **Nature conservation**: Landscape conservation is at the center, but with a focus on ecosystems and local species rather than certain scenery.
- **Yield**: Focused on efficient use of the landscape, be it through agriculture, tourism or industry.
- **Emotions**: Interest in landscape led by emotions, objects with high symbolic value are the basis of landscape constructions (e.g. village well, visually appealing houses, churches).

When such different interests in landscapes collide, "people negotiate the meaning of the contextual or environmental change as a reflection of their changing definitions of themselves" (ibid: 9/10). On the basis of the work of Greider & Garkovitch, Kühne (2018: 24ff), suggests the following central questions for investigating the societal negotiation of landscapes: What do people mean when they use the term 'landscape'? Why is this, and why is this *not*, landscape? Why and due to what needs do people construct symbolic values? When is landscape a topic, where and for whom? Who decides which landscapes are being protected? The authors underline that differences in knowledge (e.g. between superficial and in-depth or traditional and scientific knowledge), thereby influence the negotiation positions of the actors involved.

Also when researching societal processes of adaptation to climate change, understanding people's landscape constructions as well as the different meanings and values attached to those landscapes can deliver valuable insights into conflicts arising around landscape change (Köpsel et al. 2016). A social-constructivist perspective can thereby help to comprehend why certain adaptive responses are viewed as suitable by one group, but critically interfere with the self-definition of another. Thus, questioning landscape constructions "provides a vehicle for interpreting the sociocultural consequences of technological and environmental change from the diverse perspective of all participants" involved in processes of landscape management (Greider & Garkovich 1994: 14). Central questions in this context are "whose landscape is being protected, altered or exploited", what rights and responsibilities are held by the different groups of local actors, and whose perspectives can come out on top in the end (ibid: 16).

3.5 Summary of the Theoretical Approach

Despite the usefulness of the landscape perspective for understanding people's perceptions of climate change in connection with the places where its impacts occur, "relatively little attempt has been made to think through the potential for this approach in understanding the socio-political relations of climate change" (Leyshon & Geoghegan 2012: 239). To address this research gap, this study takes up Geoghegan & Leyshon's demand to "ground the idea of climate change in landscape" (ibid 2010: 293) by not only investigating how local actors make sense of a changing climate through the landscapes they manage, but by identifying differing constructions of those landscapes and their implications for physical-material adaptation activities. Moreover, I connect existing research on landscapes and climate change adaptation with the topics of landscape as a commons and the question of access to it; a topic especially relevant in the context of adaptation to an eroding coastline (cf. section 6).

As highlighted throughout this chapter, the concept of landscape in its social-constructivist understanding is very suitable for addressing individual as well as shared perceptions of local landscapes among different actors involved in processes of adaptation to climate change. Based on contemporary perspectives on the concept of landscape, I understand landscapes in this study as both physical-material and mental constructions; as both scientific concepts and everyday phenomena; as dynamic; as subjects of constant societal (re-)negotiation; and as embedded in distinct local and cultural contexts. In contrast to the concept of place, the idea of landscape thereby sets particular emphasis on the physical-material environment that shapes different societal constructions of what landscape is. It inherently connects the past, present and future, and is thus especially suitable for addressing perceptions of change; as, in this case, the change of climatic and environmental conditions (cf. sections 3.1, 3.2).

The approach to landscape research proposed (primarily) by Kühne (2008, 2018) thereby constitutes a useful analytical framework for uncovering the landscape constructions of individual persons as well as groups considering their symbolic, aesthetic, cognitive, emotional and normative facets. Applying this framework to the empirical data from the case study of Cornwall (UK) enables me to develop a detailed understanding of a) those constructions of the Cornish landscape shared across a society, b) the sub-societal landscape constructions of different actors in landscape management in the region, and c) the individual landscape constructions of professionals in landscape and environmental management on the local level. Of particular importance are thereby the different landscape constructions in existing literature (see Table 1) as well as the distinction between the landscapes of experts and lay people made by Howard (2013) and Kühne (2018) (cf. section 3.4.2). Challenging the clear distinction between natural and cultural landscapes, social-constructivist landscape research recognizes the societal constructedness not only of landscapes, but also of nature and the human-nature relationship; a factor of importance for distinguishing between the landscape constructions of different sub-societal groups in conflicts around their management and (re-)negotiation in times of change. Moreover, the conceptualization of landscapes as a commons and the question of access to them adds another theoretical layer to the discussion around different actors' approaches to climate change adaptation. Asking how different local actors construct certain landscapes, what natural and human-made elements they consider as parts of them, how they conceptualise landscape change, and what priorities for landscape management result from these constructions are therefore all central considerations for the analysis of the empirical data in this study.

To elaborate how these research objectives were approached methodically, the following chapter expands on the research design of this study as well as the choice of methods and the process of data analysis through a narrative approach.

4 Research Design and Methods

The research design of this study is that of a qualitative, explorative, non-comparative case study using a narrative approach to data analysis. It is the aim of this chapter to highlight the rationale for this approach, to provide an overview of the different methods used, and to demonstrate their suitability for the collection and analysis of the empirical data. After a broader introduction into qualitative inquiry, a short paragraph on case study research will be followed by the presentation of the qualitative methods applied during the empirical phases, namely document analysis, semi-structured and walking interviews, and explain how the data was analysed.

4.1 Qualitative Inquiry and Research Design

Although "influenced by the context, structure and timing of research", the objectives of a study usually define its methodological approach and methods of data collection (Lewis 2003: 56). In the present case, both the research questions (see section 1.2) as well as landscape research as the theoretical framework of the study (see chapter 3) imply the use of qualitative inquiry. Epistemologically mostly interpretive in nature, the aim of qualitative research is to understand "the social world through an examination of the interpretation of that world by its participants" (Bryman 2012: 380, Schnell et al. 2011). For qualitative researchers, this means taking on what Snape & Spencer (2003: 4) call an 'emic' perspective: Gathering data that enables them to see the problem in question through the eyes of those concerned with it; comprehend their interpretations of the issue; and bring to light patterns, commonalities and differences in people's perspectives (Snape & Spencer 2003, Denzin & Lincoln 2005). Qualitative research is predominantly inductive, with theories and interpretations generated from the collected data (Schnell et al. 2011, Bryman 2012, Flick 2014). Just as is the case with the strands of landscape research introduced in section 3.4, the qualitative research approach of this study takes on an ontologically constructivist perspective. Thus, "social properties are [seen as] outcomes of the interactions between individuals, rather than phenomena 'out there' and separate from those involved in its construction" (Bryman 2012: 380).

The most common methods of qualitative inquiry are various forms of interviews, document analysis and participant observation; the latter mostly used

© Springer Fachmedien Wiesbaden GmbH, part of Springer Nature 2019
V. Köpsel, *New Spaces for Climate Change*, RaumFragen: Stadt –
Region – Landschaft, https://doi.org/10.1007/978-3-658-23313-6_4

in ethnographic studies (Bryman 2012, Flick 2014). The chronology of the present research design was similar to that proposed by Bryman (2012). At the beginning a first set of research questions was formulated and a suitable case study site chosen. As the third step of the research process, Bryman (2012: 394) suggests to enter the phase of data collection. In the present case, however, large parts of the conceptual and theoretical framework were mapped out before entering the field. The analysis of the data from the first of two empirical phases lead to the development of a more refined set of questions which were addressed in a second field phase using a smaller-scale case study within the research area (see chapter 1.3.3). Analysing both sets of empirical data, a slight re-formulation of the initial research questions and a substantiation of the theoretical framework were undertaken before entering the writing-up stage of the study project.

Although many existing studies approach the topic of landscape through positivist or evaluative approaches that measure or categorise landscape perception in terms of a range of a pre-defined range of values (e.g. Jacobsen 2007, Lopéz-Martinéz 2017), this study takes the position that people's perceptions of local landscapes, changes to those landscapes and, closely related, climate change can best be comprehended by uncovering the stories they tell about these issues and the lines of argument they follow. A number of recent works in the field of landscape research support this narrative approach by applying a qualitative methodology to shed light on how people make sense of the landscapes they live and work in. For instance, Gailing & Leibenath (2015) gathered empirical data through semi-structured interviews and document analysis to study the social construction of landscapes in the German Spreewald region. Moreover, in the context of landscape and climate change Geoghegan & Leyshon (2012) assessed the ways in which farmers make sense of the changing climate through changes in their everyday landscapes by conducting in-depth interviews to uncover individual perceptions of landscapes and climate change, respectively, and the subjective values and meanings attached to these phenomena. Also using a qualitative approach, the present doctoral project applies two methods that have received limited attention to date in the context of landscape and climate change research. Firstly, local actors' perceptions of 'their' landscapes were investigated through walking interviews, and secondly different landscape constructions were assessed by identifying competing narratives about landscape, landscape change and climate change. Before outlining the methods used in the empirical phase of this study in more detail, however, the following section will give a short introduction into the use of case studies as a research design.

4.1.1 Qualitative Case Study Research

Besides comparative, retrospective and longitudinal studies, case studies are one possibility of how to design a qualitative research project (Flick 2004, Bryman 2012). A case, in this context, "connotes a spatially delimited phenomenon [...] observed at a single point in time or over some period of time" (Gerring 2007: 19). The term 'case' "is rather broadly understood here – in addition to persons, social communities (...), organisations and institutions (...) could become the subject of a case analysis" (Flick 2004: 147). Other than in studies focusing on phenomena that could be equally well researched in other contexts, the 'case' – the location or organisation itself – must stand in the centre of interest to constitute a case study research design (Bryman 2012). Thus, case study research "can further the understanding of the concept and phenomenon [...] through empirical analysis where practices [...] are examined within the context of local and regionally specific [...] challenges and cultures of practice" (Walsh & Allin 2012: 382). This focus on the particular features of the case make qualitative case studies idiographic endeavours (Bryman 2012). Their aim is not, as often the case in quantitative studies, the generalisation of the empirical results. The focus much rather lies on answering specific research questions relevant to the case study to later inductively develop a theory or theoretical framework from the case (Yin 2009, Bryman 2012). Sayer (2000) notes that the importance of the results of a case study does not depend on the extent to which they can be generalised to a larger population or number of cases. Although generalisability of the empirical findings is limited in qualitative case studies, they nonetheless provide a useful and well-accepted approach to exploring context-specific research questions using a specific example (Gerring 2007, Walsh 2010).

Yin (2012) differentiates between four distinct case study designs: single- and multiple-case designs as well as holistic and embedded. Whereas in multiple-case studies a variety of cases with different contexts are compared to each other, single-case designs concentrate on only one location, organisation or event. A holistic case study investigates only one case in each context, whereas embedded studies focus on various cases within the same context (ibid). A single-case research design is especially useful in explorative studies (Yin 2009), but can also provide a much higher level of detail in qualitative studies which hare not merely explorative in nature (Baxter & Jack 2008: 548). The present study is an explorative, holistic single-case study (comp. Sayer 2000, George & Bennett 2004, Yin 2009) focusing on Cornwall (UK) as the larger context in that the smaller-scale case study of Godrevy constitutes primary focus of the in-depth research. 'Holistic' in this context does not mean that the case under investigation is analysed in its entire complexity, but that Godrevy is the only specific case researched in the larger context of Cornwall and that no cross-case comparison is being undertaken (Yin 2009, 2012). In the words of YIN (2009), the present study ful-fils the criteria of a 'representative' or 'typical' case study by, although not an extreme or unusual case, being especially suitable for "examining key social pro-cesses" in the field of societal landscape construction and climate adaptation (Bryman 2012: 70, referring to Yin 2009).

4.1.2 Narratives in Qualitative Research

> *"The stories of who we are, which are often connected to the stories of*
> *where we live, act as a backdrop against which decisions are made. These*
> *decisions can affect both the physical and symbolic constitution of a place."*
> *(Alkon 2004: 165)*

Analysing interview data with a narrative approach and comprehending the sto-ries through which people make sense of certain phenomena has become an im-portant tool in qualitative research (Ritchie 2003, Flick 2014). The narrative ap-proach is most commonly used in the disciplines of psychology, sociology, an-thropology and ethnography, and is closely linked to a social-constructivist meth-odology (Mitchell & Egudo 2003, Chase 2005). Alkon (2004: 148), studying 'heritage narratives' in the context of soil erosion in California, understands nar-ratives about places as "emergent stor[ies] appropriated by actors that make [...] choices seem relevant and natural" (cf. introductory quote above). In this light, narratives are personal, subjective, emotional and experiential accounts of occur-rences and circumstances in people's lives, which reflect their understanding and knowledge of a phenomenon or situation (Chase 2005). The focus of identifying such stories lies thereby not on comparing different narratives with a presumed 'truth', but on uncovering storylines, motifs and values which underlie certain actors' decisions (DeSilvey 2012). From a perspective of qualitative research, bringing to light these stories provides a means for revealing "what is significant to [them] about various practices, ideas, places, and symbols" (Feldman et al. 2004: 148, referring to Young 1996). A useful distinction between stories and narratives is made by Feldman et al. (2004). The authors refer to a narrative as "the grand conception that entertains several themes over a period of time", while stories are understood as a subcategory of narratives referring to smaller themes within the larger whole (ibid: 149). A narrative can thus consist of a variety of stories which together make sense of a phenomenon in its wider context (ibid).

Previous studies have demonstrated the influence that particular narrative frames have on structuring policy debates on nature protection and environment-society relations (Greider & Garkovich 1994, Hajer 1995, Leyshon & Walker forthcoming). In the context of studying public policy, Feldman et al. (2004) find that the identification of narratives is a useful approach to comprehend how organisations and individuals communicate certain topics, create common under-standing, and shape different groups' decision-making processes. The authors emphasize that "one can expose implicit understandings in narrative, reason, or representational practices, without also claiming that this is the only way to in-terpret a narrative" (ibid: 151). In her work on 'participatory storylines', Hen-dricks (2005) under-stands a narrative not as "a plot with a beginning, middle and an end", but as "metaphor[s] for a set of coherent ideas, which circulate around a given issue" and constitute specific lines of argument followed by cer-tain actors (ibid: 4). A narrative approach is most commonly applied for three reasons: Investigating how people represent events in their lives through story-telling; Analysing the structure and textual coherence through which certain mes-sages are brought across; and understanding the functions of stories in setting particular contexts, for example for policy processes, and the consequences these stories have. The latter is the aim of this study: to uncover what narratives exist about the Cornish landscape and climate change in general and Godrevy in par-ticular, and what the implications of these narratives are for local negotiation processes in the context of climate adaptation.

While not an uncommon approach in the field of human geography and place research (see e.g. Johnstone 1990, Buciek et al. 2006, While & Short 2011), the identification of narratives in research on individual understandings of cli-mate change, and especially adaptation to it, is still relatively unexplored. Adger et al. (2013) suggest that one reason for the absence of purely qualitative methods in climate adaptation research is that "these methods do not sit comfortably with the quantitative approaches prevalent in other social and natural science on cli-mate change" (ibid: 112). Exceptions, however, exist: In the mid-1990s van Eeten (1997), researching policy narratives about dike improvements at the Dutch North Sea, identified a number of stories and counter-stories about the strengthening of dikes and argues "that the controversy [*about the suitable pol-icy*] issues from [...] the 'moralizing impulse' of the stories" (Wagenaar 2014: 220, referring to van Eeten 1997). Using the example of delta management in the Netherlands, Verduijn et al. (2012) show that communicating climate change and adaptation through 'crisis narratives' can create a sense of urgency and a push for collective action towards the implementation of adaptation measures.

In dialogue with different actors in landscape management Cornwall (UK), Leyshon & Walker (forthcoming) use the term 'framing' when referring to dif-ferent storylines about the management of landscapes in their local context. The authors argue that different narrative frames around a particular problem deci-sively shape the outcomes of environmental management processes. If more than one such narrative frame exist around a certain issue, and if these frames vary between different actors in landscape management, I argue, they can have direct implications for the implementation of physical-material climate adaptation measures at the local level and hold the potential for conflict. The following section will expand on the concrete use of narratives in the present study. However, before moving to this section, the often seemingly synonymous use of the terms storylines, frames and narratives calls for a short digression into the field of terminology.

The Understanding of Narratives in this Study

As Leyshon & Walker (forthcoming) suggest, one way of looking at people's storylines around certain issues is the concept of frames. Even if originating in different schools of thought, the concepts of narratives and frames have similar implications in practice. Deriving from psychology and sociology, frame analysis has been increasingly used in the fields of environmental management and interpretative policy analysis (Rein & Schön 1993, Hajer 2003, Lindahl et al. 2014, Leyshon & Walker forthcoming). Leyshon & Walker (forthcoming) highlight that frames, like narratives, are influenced by actors' personal experiences, organisational backgrounds and prevalent societal institutions. The authors argue that how certain topics in environmental management are "conceptualized and defined in the first instance" are important determinants for the goals of management and the course of action chosen (ibid: 2). A classical definition of framing was made by Entman (1993: 52) who argues that "[t]o frame is to select some aspects of a perceived reality and make them more salient in a communicating text, in such a way as to promote a particular problem definition, causal interpretation, moral evaluation, and/or treatment recommendation". The definition of framing is thus not very different from that of narratives. However, there is a decisive difference: in contrast to narratives, Entman (1993) suggests that the act of framing is firstly an intentional, secondly a strategic, and thirdly a normative process (see also Verduijn et al. 2012).

To further distinguish between the concepts of frames and narratives, it is helpful to specify the etymology of each concept. In general terms, a frame is defined as "[a] rigid structure that surrounds something such as a picture" or "[a] basic structure that underlies or supports a system, concept, or text" (Oxford Dictionary Online 2016). As this definition suggests, the concept of framing is directed at differentiating a debate or topic and giving it clearly pronounced, static boundaries distinguishable from other frames by means of wording and structure of arguments. A narrative, on the other hand, may be defined as "[a] spoken or written account of connected events; a story" as well as "[a] representation of a particular situation or process in such a way as to reflect or conform to an overarching set of aims or values" (Oxford Dictionary Online 2017a). Narratives thus imply a strong emphasis on the development of an overarching plot over time and on subjective values and meanings, thus being especially useful when researching individual perceptions of processes, in this case the changing of landscapes and the climate.

In summary, a narrative approach is most commonly used in existing literature with two aims: On the one hand, and understood in the sense of frames, it is applied to uncover intentionally constructed stories put together by different actors to pursue a certain goal or agenda (comp. Verduijn et al. 2012, Leyshon & Walker forthcoming). On the other hand, it is used to understand the ways in which individuals make sense of events or phenomena by relating them to their personal life worlds, experiences and social contexts (Alkon 2004, Chase 2005). From the viewpoint of landscape research, such narratives form a bridge between individual perceptions of landscapes and collective constructions shared across a society (see section 3.4.1). In the context of the present study, narratives were chosen as an analytical approach to the interview data for three reasons related to the research questions and theoretical embedding: Firstly, a narrative approach responds to the call for a more values-based, subjective perspective on processes of climate change adaptation by centring different actors' perceptions, values, and the internal lines of arguments connecting different parts of their larger storyline.

Secondly, the focus of narratives on the temporality of the recounted events accommodates well the objective of uncovering how local actors in landscape management make sense of past, current and future changes to the landscape both through climate change as well as through human-made interventions.

Thirdly, viewing each interview as an individual story about landscape, landscape change and climate change brings to light not only how the landscape itself is perceived as a phenomenon, but also how the interviewees see themselves in relation to it. Applying Hendricks' (2005) definition, this study understands narratives as overarching, coherent lines of argument through which different actors make sense in diverse ways–relating to their professional knowledge *and* personal experiences–of the Cornish landscape and climate change. A narrative approach therefore contributes to transcending the divide between experts and locals by revealing how professionals make sense of landscapes and climate change in their daily work and against the background of their own experiences (cf. section 2.3). These narratives are, however, not given, coherent storylines that exist naturally around an issue and are formulated as such by the interviewees. The identification of narratives is here understood as a twofold interpretive process in which firstly the study participants offer their own interpretation of the investigated phenomena to the researcher in a non-linear, not necessarily coherent way by answering to the interview questions. Secondly, by analysing the interview data and grouping them into stories and narratives with similarities and differences, the researcher interprets these interpretations on the basis of theoretical concepts and research objectives (Flick 2014).

Against the background of the research questions of this study (cf. section 1.2) and the utility of a narrative approach for understanding the societal dynamics of climate adaptation (see above), the next chapter will outline the methods for data collection and analysis applied in the context of this study as well as highlight the steps taken to identify different narratives from the empirical data.

4.2 Data Collection and Analysis

The data collection for this study was conducted in mid and west Cornwall (UK) during two empirical phases. The first field phase took place between September and November 2015 and focused on exploring the general connections between different perceptions of the Cornish landscapes, climate change, and adaptation among local actors in landscape management. The second field phase was conducted between late April and early June 2016 at the smaller case study site of Godrevy, west Cornwall, to investigate in greater depth the implications of different landscape constructions for specific processes of climate change adaptation at the local level. The following sections give a literature-based overview on the methods used for data collection before outlining their practical application during the empirical process of this study and the corresponding means of data analysis.

4.2.1 Document Analysis, Semi-Structured and Walking Interviews

According to Feldman et al. (2004), both written documents and interviews are suitable means of collecting and identifying narrative data. The findings of this study are based on both naturally occurring data gathered through the analysis of strategy and policy documents as well as newly generated data from semi-structured and walking interviews.

Document Analysis
Documents, either officially published or private, are, as Lewis (2003) terms it, 'naturally occurring' sources of data in qualitative research. In addition to private notes like diary entries or letters, such documents can be media reports, government publications or publicity materials, minutes of meetings, formal letters or financial accounts (Ritchie 2003: 35). Bowen (2009: 27) highlights that "document analysis requires that data be examined and interpreted in order to elicit meaning, gain understanding, and develop empirical knowledge". It is often used in qualitative research to facilitate a triangulation with other methods of data

collection. The aim of such combination of methods can, for example, be the comparison of the ways in which an issue is framed and portrayed in an organisation's policy documents with how actors and stakeholders interpret it in practice (ibid). Document analysis is thus especially "applicable to qualitative case studies" (Yin 2009, Bowen 2009: 29). Bowen (2009) highlights five useful applications of document analysis in qualitative inquiry: Firstly, documents can provide important information about the context and history of the case under investigation and thus be a basis for further data collection. Secondly, the analysis of documents can hint to topics and questions that are important to consider in the interview phase. Thirdly, the content of the analysed documents can serve as additional data during the research process with regard to different actors' viewpoints and lines of argument. Fourthly, changes and developments in the portrayal and communication of certain topics, or in the perspective of certain actors, can be traced through the analysis of documents across a wider span of time. Lastly, document analysis can be used as a means of cross-checking data obtained through interviews or observations in the field.

Semi-Structured Interviews

Interviews are the most common method of qualitative inquiry (Flick 2004, Denzin & Lincoln 2005, Schnell et al. 2011, Bryman 2012). They have several advantages for gathering high-quality and in-depth data: Being face-to-face and bilateral in nature, interviews can create deep insights into the perspectives, experiences and knowledge of the interviewees. They moreover provide the opportunity for clarifications as well as further (and spontaneous) questions about the topic of interest and make it possible to see the phenomena in focus through the eyes of the participants (Ritchie 2003, Bryman 2012).

From an interpretive perspective, interviewing is always a collaborative effort shared between the interviewer and the interviewee(s) (Fontana & Frey 2005). An interview is seen as an active process during which a "contextually bound and mutually created story" is developed through a dialogue between researcher and participant (Holstein & Gubrium 1995, Fontana & Frey 2005: 696). Common forms of social scientific interviews are the structured, semi-structured and narrative interview; the latter being a very open form of interviewing with a decisive part of the talking done by the interviewee and only very few questions posed throughout the process (Fontana & Frey 2005, Bryman 2012). Semi-structured interviews are oriented by a pre-formulated list of questions, the interview guide (Schnell et al. 2011, Bryman 2012). Although this study applies a narrative approach to data analysis (see chapters 4.1.2, 4.2.3), the interviews conducted during the empirical phase were semi-structured in nature and followed such an interview guide. Whereas structured interviews suggest a number of pre-formulated response categories, semi-structured interviews consist of open questions and leave room for flexibility during the interview process (Fontana & Frey 2005). Thus, they are "flexible, responding to the direction in which interviewees take the interview and perhaps [*lead to*] adjusting the emphases in the research as a result of significant issues that emerge in the course of the interviews" (Bryman 2012: 470). However, since qualitative studies do not aim at producing representative or generalisable results and instead focus on portraying the diverse perspectives on the issue under investigation, the overall number of respondents is generally relatively small (Snape & Spencer 2003: 470).

Walking Interviews

Although also in the case of classic, sedentary interviews "researchers 'go out' into the context of the 'field' […] to […] become cognisant of the context-specific practices that form the meanings and knowledges in question" (Anderson 2004: 255), the method of walking or 'go-along' interviews (Capriano 2009) takes on a special role in researching places and people's relations to them. Walking interviews belong to the field of the newly developing 'mobile methods' which have gained popularity in recent years especially within ethnography and human geography (for details on the 'new mobilities' paradigm see Sheller & Urry 2006). Possible applications of walking interviews range from asking participants to present their beloved places to following practitioners around at their (outside) places of work, or structured tours at specific locations (Evans & Jones 2011). Such interviews in motion are often, but not necessarily, coupled with technologies such as mobile GIS or GPS trackers (e.g. ibid 2011).

By taking an interview outside and setting it in motion, Evans & Jones (2011: 849) find that "the data generated through walking interviews [is] profoundly informed by the landscapes in which they take place, emphasising the importance of environmental features in shaping discussions" and for "the production of rich place narratives". In the words of Anderson (2004: 245), "'talking whilst walking' can harness place as an active trigger to prompt knowledge recollection and production" and thus create an additional layer of information during the interview situation and in the data. Such joint being in the landscape can facilitate spontaneous dialogue with the interviewees about the material elements of the landscape and changes to them, but also provides the possibility of addressing topics difficult to discuss without a tangible example at hand. Walking interviews are most commonly used in existing literature to address matters of attachment or personal accounts of places (Moles 2008). This study, however, explores the suitability of walking interviews for transcending the dichotomy between the interviewees' roles as experts and their personal experiences with and attachments to the landscape by addressing both their professional and affective relationships with the place under investigation.

This dynamic form of interviewing moves into focus the places that stand in the centre of research. Being in the location under investigation with the interviewees is useful in "prompting and recalling personal knowledge" (Anderson: 257) and changes the ways in which interviewees relate to both those places and the activities they perform in them. In their work on farmers' accounts of climate change in Cornwall (UK), Geoghegan & Leyshon (2012) find that interviewing their participants in-situ, on their farms and fields, enables them to gain valuable insights into how the changing climate is experienced and responded to by locals through their everyday practices. Being outside and walking, as Evans & Jones (2011: 850, referring to Solnit 2001) argue, "has long been considered a more intimate way to engage with landscape that can offer privileged insights into both place and self". This means being able to uncover interviewees' personal accounts of the landscape and their relationships with it through what Ingold (2004) refers to as the practice of 'dwelling' (cf. section 3.1), experiencing a place through the everyday bodily encounter with it (see also Ward Thompson 2013). The method of walking interviews thus 'grounds' people's understandings of the landscape and changes to it by placing the interview conversations directly into the, as Kühne (2008, 2018) terms it, physical basis for their landscape constructions and connecting them to the materiality of the landscape.

The effectiveness of this approach for addressing the research objective of transcending the dichotomy between the role of experts and their personal relationships with the places they manage will be discussed in detail in the subsequent empirical chapters as well as in section 7.3. Now first of all I will outline the methods used for data collection during this study as well as their application during the two field phases in Cornwall.

4.2.2 Application of Methods

The following paragraphs will outline in detail how the methods of document analysis, semi-structured and walking interviews were carried out during the two field phases before discussing the process of data analysis.

Entering the Field through Document Analysis

In this study, the analysis of documents was undertaken for two purposes. Firstly, as suggested by Bowen (2009) (see section 4.2.1), online publications by different organisations involved in landscape management and climate adaptation in Cornwall served as an important instrument in contextualizing the case study and informing the empirical fieldwork by providing an overview on the relevant local actors in environmental and landscape management, their rights and duties, and their perspectives on landscape and climate change in the area. The analysis of documents before entering the field for the first time was thus an important step in the research process. Moreover, the large majority of official documents published by government or organisations, follow certain logics and lines of argument aiming at portraying a phenomenon in a specific way – in the words of Leyshon & Walker (forthcoming), they are following specific narratives frames. Uncovering these different narrative frames circling around landscape, landscape change, climate change and adaptation within publications by the relevant organisations in Cornwall was the second purpose of analysing documents in the context of this study.

Two sets of document analyses were carried out in the course of this study both as a means of entering the field as well as to investigate what different constructions of the landscape and climate change become evident from existing documents by the organisations involved. Documents analysed were newspaper articles, policy documents by Cornwall Council, policies and management plans by organisations such as the Environment Agency, Natural England and the Cornwall AONB Partnership, as well as related websites. Particularly important documents by Cornwall Council turned out to be the 'Cornwall Local Plan 2010-2030' with its sections about landscape management and climate change as well as the Council's websites on landscape, heritage, and climate change (Cornwall Council 2009, 2011a, 2014b, 2015, 2016b, 2017). To gain an overview of current activities around adaptation to sea and river flooding in Cornwall, a number of management plans by the Environment Agency provided comprehensive insights into related activities in the area (e.g. Environment Agency 2012b). Regarding different landscape designations and related organisations, the Cornwall Council Interactive Map held a multitude of information (Cornwall Council 2016a). All these documents were read closely with regards to a number of guiding questions:

- What actors in Cornwall are involved in landscape management on the one hand and climate adaptation management on the other?
- How do these actors portray the Cornish landscapes and changes to them?
- What main effects of climate change on Cornwall are presented and who is responsible for addressing them?
- What adaptation activities are suggested as necessary by whom, and which information do the documents hold regarding their implementation?

On the one hand, the results of this document analysis served as an orientation for establishing contact with potential interview partners for the first field phase. On the other, the data collected provided some important factual knowledge directly relevant to the research objectives of this study (cf. section 1.2).

The second set of document analysis was conducted with focus on the smaller-scale case study of Godrevy. After gaining an overview of the organisations involved in landscape management at the site through online research (especially using the Cornwall Interactive Map), policy documents and management plans of selected relevant organisations particularly referring to Godrevy were identified and read closely. Especially relevant for this second document analysis were the web presence of the National Trust about Godrevy (National Trust 2016b), their 'Shifting Shores' coastal management policy (National Trust 2017e), as well as the AONB's 'Management Plan 2011-2016' (Cornwall AONB Partnership 2011b). Specific questions addressed were similar to the guiding questions from the first set of document analysis, but with special emphasis on Godrevy:

- How do the different actors involved in landscape management at Godrevy construct the area's landscape and changes to it? What are their management priorities?
- How is climate change portrayed by the different actors and which impacts on Godrevy are presented in the documents?
- Which new challenges are outlined for landscape management at Godrevy due to climate change and what adaptation measures are proposed?

Additional documents for this second analysis were provided by a number of interview partners in the form of handouts, hardcopies of policy documents, brochures and drafts for information boards. For a complete list of documents analysed during the first and second field phases, please see Appendix 2.

Capturing Individual Views with Semi-Structured and Walking Interviews
Altogether, 31 semi-structured and walking interviews provided the main empirical data for this study. These interviews were conducted with professionals from the fields of environmental and landscape management responsible both for the larger Cornwall area (1st field phase) and the Godrevy headland (2nd field phase). As expanded on in section 2.4, this study applies a distinct understanding of the role of local professionals aiming at transcending the dichotomy between scientific expert thinking on the one hand and local, experiential knowledge and personal attachments on the other. In this context, the actors in landscape management interviewed in the course of this project are therefore viewed as locally embedded individuals occupying a hybrid position between expert and local, personal and professional knowledge. In the following, I will outline how the methods of semi-structured and walking interviews were applied to uncover these actors' perceptions of the Cornish landscapes, changes to them, and climate change in their respective local contexts.

First Field Phase – Semi-Structured Interviews
The field research was conducted in form of a three-month long stay in Cornwall and the placement as a guest researcher at the Department of Geography on University of Exeter's Cornwall Campus. This longer-term stay in the research area enabled me to get familiar with Cornwall as a place and the local debates about climate change and adaptation on the one hand, and made it possible to arrange interview appointments on short notice on the other. Potential interview partners for the first field phase were identified through what Richie et al. (2003) term a 'criterion based' or 'purposive' sampling strategy. In this strategy, interview partners "are chosen because they have particular features or characteristics which will enable detailed exploration and understanding of the central themes and puzzles which the researcher wishes to study" (ibid: 78). However, the aim of this study's sampling strategy was not to explicitly address organisations from pre-defined fields (e.g. set societal groups such as nature protection, government, communities, etc.) as often done in landscape-related studies (comp. Gailing 2012b). Following a grounded theory-lead approach (cf. section 4.2.3), the goal was much rather to interview representatives of all actor groups involved in landscape management in Cornwall detached from a priori identified categories.

On the basis of the document analysis conducted previously and further online research, a number of key organisations in landscape and adaptation management in Cornwall were identified; among others Cornwall Council, the Environment Agency, the National Trust, Natural England, and the Area of Outstanding Beauty (AONB) Partnership (see also section 1.3.1). The interviews taking place during the first field phase were sedentary, semi-structured interviews. At the end of each interview, the participant was asked for further potential interview partners in the field of landscape management and climate adaptation, and often contact details were provided for further inquiries enabling additional sampling through a snowball approach (see Richie et al. 2003: 94, Bryman 2012: 424). A complete, anonymised list of interview partners, dates, locations and durations of the interviews can be found in Appendix 3.

The interview guide used during the first field phase was developed based on the theoretical framework of this study and the research questions guiding the data collection (Schnell et al. 2011) (see section 1.2). Minor adaptations were made to each interview guide depending on the affilitaion and role of the interview partner (see Table 3). Each interview guide did, however, cover the same topics in the same order: (1) personal and professional background of the interviewee; role and duties in organisation; (2) perceptions of the Cornish landscapes; interviewee's relationships with the landscape; (3) perceived landscape changes in Cornwall; actors influencing landscape change in Cornwall; (4) interviewee's understanding of climate change and impacts on Cornwall; professional involvement with and personal affectedness by climate change; (5) adaptation activities in Cornwall, especially by interviewee's organisation; relevance of landscape-related issues in adaptation planning; responsibilities and negotiation around climate adaptation measures; conflicts around adaptation in Cornwall; (6) anticipated landscape changes and ideal management of landscapes under climate change in the future. The interview guide used during the first empirical phase can be found in Appendix 4. The ways in which the data gathered through the first empirical phase was analysed will be laid out in the following in section 4.2.3.

Table 3:   Interview partners sorted by actor groups, 1st Field Phase

Actor Group	Organisation	Abbreviation
Government Bodies	Cornwall Council (3)	Counc
	Cornwall Councillors (2)	Counc'lor
	Parish Councillors (2)	Par'lor
Environmental Management Organisations	Environment Agency (2)	EA
	Natural England (1)	NE
	Cornwall Wildlife Trust (1)	CWT
	Upstream Thinking Project (1)	UT
Landscape Mgmt. Organisations	National Trust (2)	NT
	AONB Partnership (1)	AONB
Themed Regional Organisations	Community Flood Forum (1)	CCFF
	Climate Vision (1)	CV
	Visit Cornwall (1)	VC
	Cornwall Sustainable Tourism Network (1)	CoaST

Source: own compilation

Second Field Phase – Walking Interviews at Godrevy
To investigate the concrete implications of landscape constructions on local climate change adaptation activities, further interviews were conducted with actors involved in landscape management at Godrevy during a second field phase from April to June 2016. As outlined in section 4.1, the qualitative research process is adaptable and remains flexible towards new topics of inquiry emerging from the interviews already conducted (Bryman 2012). In tune with a grounded theory-led approach to data analysis (see section 4.2.3), the choice of Godrevy as a second case study was based on the interviewees' accounts of current climate adaptation activities in Cornwall and their framing of Godrevy as a case of conflict related to climate change. It moreover proved to be a suitable case study to

address this project's research questions due to the high number of actors from landscape management involved in decision-making at Godrevy (cf. section 6.1). With regards to the first research question of this thesis, namely to investigate both the professional expertise and the personal relationships of local actors with the landscape, it was important in the case of Godrevy to choose a method that addresses questions regarding both the interviewees' expert knowledge as well as their emotional attachments. Therefore, although following an interview guide similar to the first empirical phase (see below), the second round of interviews was approached methodically by conducting so-called walking interviews with the participants. The inclusion of walking interviews as a method was explorative in the course of this doctoral project in two regards. On the one hand, it was a methodological aim of the present study to explore the applicability of walking interviews for enriching the data gathered during the interview conversations with visual and acoustic cues from the physical environment.

Twelve such walking interviews were conducted during the second field phase at Godrevy. The participants of the second round of interviews were also identified and approached on the basis of a purposive sampling strategy (Richie et al. 2003). Through the Cornwall Interactive Map (Cornwall Council 2016a), the National Trust and AONB websites (National Trust 2016b, Cornwall AONB Partnership 2016a) as well as further online research, relevant organisations involved in landscape management at Godrevy were identified and contacted with requests for interviews. All participants who were approached with requests to meet at the site and conduct such an interview responded positively to this special form of research.

Table 4:   Overview Interview Partners, 2nd Field Phase

Actor Group	Organisation	Abbreviation
Landscape Mgmt. Organisations	National Trust (2)	NT
	AONB Partnership (1)	AONB
Environmental Mgmt. Organisations	Natural England (2)	NE
	Cornwall Wildlife Trust (1)	CWT
Themed Regional Organisations	South West Coast Path Assoc. (1)	SWCPA
	Visit Cornwall (1)	VC
	Cornwall Seal Group (1)	CSG
Local Actors	Towans Partnership (2)	TP
	Tenant Farmer (1)	TF
	Local Surf School (1)	SS
	Local Resident (1)	R

Source: own complication

Key organisations managing Godrevy turned out to be the National Trust (from here on 'the Trust' or NT) as the owner of the site, the AONB Partnership, Natural England and Cornwall Wildlife Trust (cf. section 6.1). Moreover, representatives of the neighbouring Towans Partnership and local actors such as the owner of the local surf school and the tenant farmer were interviewed. The Parish Council of the neighbouring village of Gwithian did not officially agree to an interview; however, a local resident who is member of this Council was interviewed in his private role. For an anonymised list of interviewees at Godrevy, see as well Appendix 3. The interview guide used during the walking interviews at Godrevy was semi-structured in nature and similar to that used during the first empirical phase. Some questions were adjusted to be more specific to the situation at Godrevy, addressing the coastal erosion and the negotiations about climate adaptation (see Appendix 5). Unlike in many other forms of walking interviews (e.g. Evans & Jones 2011), the interviews at Godrevy were not technology-aided aside from a recording device and a clip-on microphone used to tape the conversations.

Evans & Jones (2011) distinguish between different forms of walking interviews based on the location of the interview as well as who (interviewer or interviewee) chooses the route walked. The interviews conducted at Godrevy were what the authors, referring to Clark & Emmel (2008), term 'participatory walking interviews'. Since the interview site was familiar to all interviewees, every participant was asked at the beginning of the interview to choose the route they preferred to walk while speaking about Godrevy and changes to its landscape. As the system of paths at Godrevy is not extensive, however, the possibilities of routes to choose were limited, resulting in similar routes in each interview.

Two of the interviews conducted during the second field phase were, as Ritchie (2003: 37) term them, 'paired interviews' with two participants, being in dialogue with both interviewees at the same time and giving them room for discussion with each other. The walking interviews lasted between 50 minutes and 1.5 hours and were followed by a phase of detailed note-taking by the researcher regarding the route walked, the weather conditions, the crowdedness of the place, the general atmosphere of the interview as well as the points on route at which the interviewees stopped and the conversation was held focusing a particular view or discussing elements of the landscape in more detail (see Figure 7 and Appendix 6). Due to the relatively small spatial extent of the Godrevy headland, the method of walking interviews proved to be especially suitable to uncover the different perspectives on the local landscape. This dynamic and in-situ way of interviewing added a rich layer of information to the conversations when the participants took cues from the physical landscape, which triggered their memories or prompted them to address certain issues with relation to what they saw or heard while walking across the headland.

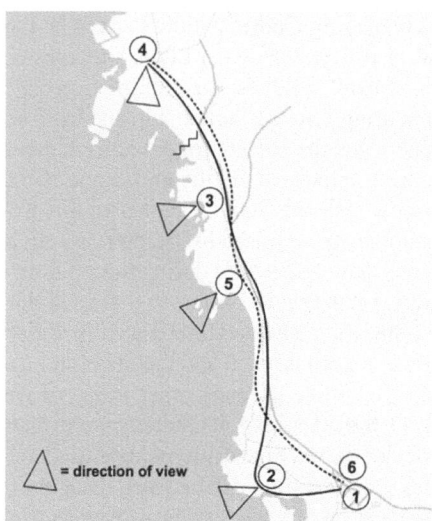

Figure 7:   Sketched path of a walking interview at Godrevy
Source: own sketch based on maps.google.com

Aside from the data about different perceptions of the Godrevy landscape and changes to it, climate change effects on the site and the proposed adaptation measures, the walking interviews provided deep insights into how the interviewees related to the landscape and which personal connections they had with it. The results from both empirical phases will be presented and discussed in length in chapters 5 to 7. Beforehand, however, the following subchapter outlines how the interview data were analysed as well as which approach was applied to identify narratives within the accounts made by the participants.

4.2.3   The Process of Data Analysis

> *"There is a difference between an empty head and an open mind."*
> *(Dey 1999: 151)*

Spencer et al. (2003: 199) find that "analysis is a challenging and exciting stage of the qualitative research process. It requires a mix of creativity and systematic searching, a blend of inspiration and diligent detection". Even if qualitative research and data analysis are inductive processes with scope for interpretive freedom on the side of the researcher there are, nonetheless, standard procedures that can be applied (Denzin & Lincoln 2005, Bryman 2012). This study follows the grounded theory paradigm, an approach developed during the 1960s by Glaser

& Strauss (1967) and refined by Corbin (1998) and Dey (1999). Analysing data from a grounded theory perspective means to set aside any pre-assumptions about possible results, and develop codes and categories *from* the interview data through grounded coding (Charmaz 2006, Bryman 2012). The quote opening this subchapter aptly summarizes the researcher's ideal state of mind during the analysis process: whereas it is neither possible nor practical to approach the collected data with an entirely empty mind, as suggested in the early stages of grounded theory, the key idea of this approach is much rather to set aside any expectations to the data or pre-formulated categories based on the research questions and theoretical framework, and firstly take the data for what is really *said* by the interviewees (Dey 1999, Charmaz 2006). This grounded, inductive approach to data analysis has become one of the most common approaches in qualitative research (Bryman 2012). The following sub-sections will outline how to approach the coding process in more detail and how to identify narratives from interviews transcript from a grounded-theory led perspective. Subsequently, I outline how these steps were undertaken during the present study.

Transcription of the Interview Recordings
Transcription in qualitative research is the process of transforming recorded interview data into written text (Ritchie & Lewis 2003, Bryman 2012, Flick 2014). The interviews from the first empirical phase were transcribed verbatim using Microsoft Word and VLC Player. Irregularities within the interviewees' responses such as clearing one's throat or hesitations were thereby left out of the transcription. This is legitimate, Flick (2009: 300) argues, if the goal of data analysis is the interpretation of the context and no psychological or linguistic analysis. Although "precise transcription of data absorbs time and energy" (ibid: 300), the process of re-listening and typing up the interview recordings was useful to become more familiar with the contents of the interviews. Charmaz (2006: 47) argues that the researcher relives the interview situation through the transcription process as "[w]e interact with [*the participants*] again many times over through studying their statements and observed actions and re-envisioning the scenes in which we know them". The process of typing up interview recordings thus not only serves the purpose of bringing them into written form, but also as a first step of becoming more aware of and familiar with the contents of the data.

Whereas all interview recordings from the first field phase were transcribed verbatim, those from the second phase were not transcribed entirely for two reasons: Firstly, a complete transcription of interview recordings is time-consuming and would have led to extensive time pressure. Secondly and more importantly, the second round of interviews served as an in-depth study of the results from the first empirical phase, namely the different landscape narratives and their implications for climate adaptation. After the analysis of the first set of interviews, I was already familiar enough with these topics to depart from a complete transcription and analyse the data by re-listening to the recordings and directly retrieving and transcribing those sections that regarded the key issues identified during the first analytical phase. This procedure is supported by Flick (2009: 300, referring to Strauss 1987) who argues that "[i]t seems more reasonable to transcribe only as much and only as exactly as is required by the research question".

Coding and Narrative Identification
As outlined in section 4.1.2, one aim of data analysis was to identify narratives about the Cornish landscape and climate change. The identification of narratives from qualitative data is an interpretive process with no strictly set rules. However, a good orientation is provided by Feldman et al. (2004) who suggest a useful three-step approach to narrative analysis. The authors suggest to firstly analyse each interview with regards to the smaller *sub-stories* it consists of; in the case of this study, these are sub-stories about what the landscape is; understandings of climate change; and approaches to adaptation. Secondly, the *broader line of arguments* stretching through each interview and connecting the different sub-stories is identified. The last step consists of grouping together interviews that followed similar lines of argument and forming what the authors term '*encompassing narratives*' (ibid). To identify 'sub-stories', a grounded theory-led coding approach was applied in this study. Coding is "the key process in grounded theory, whereby data are broken down into component parts, which are given names" (Bryman 2012: 568). In the words of Snape & Spencer (2003: 4) this means that the focus of data analysis was the identification of "emergent categories and theories from the data rather than imposing a priori categories and ideas". Charmaz (2006) proposes a two-step coding process, the first step of which develops 'initial codes' closely to the text, whereas the second step of 'focused coding' abstracts from the transcript and initial codes are being grouped into larger categories (see Figure 8) (for list of emergent codes and categories see Appendix 7). In accordance with the paradigm of interpretive interview analysis, it is thereby important to stay open for newly emerging themes or sub-stories in the data that had not been anticipated beforehand (Charmaz 2006, Bryman 2012).

As common in the process of qualitative analysis, the data was initially approached with the software MaxQDA. MaxQDA is a computer-based aid to coding and analysing interview data with special emphasis on creating a so-called 'code tree' with overarching and sub-codes (MaxQDA 2016). After coding a small number of interview transcripts from the first field phase, however, the process of software-aided coding proved unsuitable for the identification of stories and narratives within the interview data.

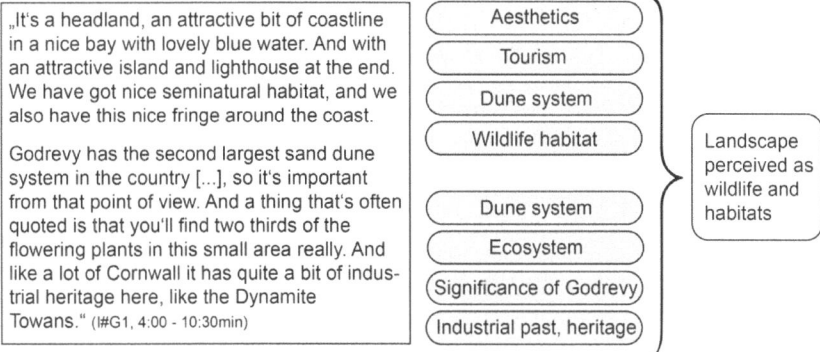

Figure 8: Example of initial and focused coding
Source: own interview transcript and coding

Spencer et al. (2003: 211) argue that "[t]oo many code and retrieve packages fragment the data to the point that the overall narrative is lost and linkages between different aspects of an individual case or story are difficult or impossible to re-create", thus expressing exactly what turned out to make the use of MaxQDA so problematic in this study. Instead, emerging themes and sub-stories were identified through iterative readings of the transcripts, manual coding, and the creation of an overview of the emerging sub-stories in an Excel sheet. Within these sub-stories, the focus of analysis was laid on characteristic wordings and phenomena addressed, particular feelings expressed as well as specific argumentations (see Figure 9).

The interview data from the second round of interviews were analysed based on the results from the first empirical phase. While still being open for newly emerging topics within the data, the larger themes identified from the first set of interviews as well as the process of narrative identification describes above served as guidelines for the analysis of the Godrevy data. However, a number of new themes and stories emerged from this second set of focused in more detail on the specific processes in the smaller-scale case study. One the other hand, and from a methodical point of view, this new emergence of themes and stories highlights the importance of keeping an open mind during data analysis, even if pre-formulated categories already exist from a previous field phase. During the data analysis, each interview was given an acronym for referencing in the empirical chapter. An overview of all interviewees with according abbreviations for the quotes in chapters 5 and 6 can be found in the table of Appendix 3.

#	Organization	What is Landscape?	Characteristics
1	Natural England	"Farmed. Reasonably wooded. Beautiful coast. Deep estuaries. Rivers that come in. Seasonal activity with sailing and people walking on the coast" (1) / "The **maritime influence** is huge." (2)	"So Cornwall has got these **contrasts,** which is really nice. You can just pick and choose." (2); "And if you go down to West Cornwall it's **different** again." (3)
1	Natural England	"It [*landscape*] is very **important to me.** I live in a farmhouse down the track on my own land" (2)	"And I suppose it's also very **remote.** Meetings...will happen at London. And it's five hours to get there." (1)

Figure 9: Excel Sheet with Sub-Story Analysis
Source: own figure

4.3 Summary of the Methodical Procedure

This study applies a qualitative, explorative, non-comparative case study approach from a threefold methodical angle through document analysis, semi-structured and walking interviews. Each of these methods served a specific purpose in investigating different landscape constructions among actors in landscape management in Cornwall and different approaches to climate change adaptation. The document analysis served as an important entry point to the field in general, but also a tool for identifying relevant organisations and potential interview partners. The semi-structured interviews with staff of landscape management organisations were an important step to lay open the co-existing narratives about the Cornish landscapes and their implications for climate adaptation; thus establishing the larger context of the smaller-scale case study of Godrevy. The third method of walking interviews proved particularly suitable in the Godrevy case to break down the expert-local dichotomy (cf. section 7.1.2) and was a valuable addition to the sedentary interviews, making the use of walking interviews a crucial part of the research design of this study.

After having outlined the theoretical framework, research design and methodological procedure of this study, the subsequent chapters are dedicated to the presentation and interpretation of the findings of this study. Chapter 5 sets out the results from the first empirical phase focusing on different landscape constructions and their implications for climate adaptation in a broader Cornwall context. In chapter 6, I focus on the case study of Godrevy to show how the narratives identified from the first set of interviews play out in a specific, smaller-scale context and in which debates around climate change adaptation they manifest among actors in landscape management on the local level.

5 Landscape Narratives in Cornwall and their Implications for Climate Change Adaptation

During the first field phase of this study, a number of shared and individual constructions of the Cornish landscapes were identified among actors in landscape and environmental management that I will present throughout this chapter. After expanding on the collectively shared perceptions of Cornwall's landscapes (5.1), I introduce what I term the official policy narrative; the way in which Cornwall Council portrays the county's landscapes and the impacts of climate change on them (section 5.2). I then present four different narratives that stand in contrast to this official policy perspective: the Cornish landscapes as (1) human-environment interaction; (2) natural systems; (3) visual beauty; and (4) functional spaces of production (section 5.3). These different narratives come with particular understandings of climate change and of how the changing climate impacts on Cornwall's landscapes. They therewith have distinct implications also for processes of adaptation to those changes; implications that will be highlighted in section 5.4. at the end of this chapter.

5.1 Common ground: Collective Landscape Constructions

As hinted to above, distinct narratives about landscape and climate change unfold throughout the interviews with local actors in Cornwall. At the same time, however, all interviewees have one thing in common: they collectively share a high appreciation of the Cornish landscapes as visually attractive and express strong emotional bonds to Cornwall as a place. Thus, the importance of Cornwall's landscapes for its regional identity that I have already argued in section 1.3.1, is mirrored in all dialogues with actors in local landscape management. Notable commonalities in the interviewees' choice of words indicate a shared prevalent discourse in the region about which landscape elements are viewed as iconic for Cornwall, and which serve as a societally shared basis for the different narratives that exist about the area's landscapes.

5.1.1 Coastal, rugged, beautiful: Shared Landscape Perceptions

Regardless of their organisational affiliation, a shared understanding exists among all interview participants about which landscape features and characteristics are typical for Cornwall. These agreed-upon characteristics are primarily

© Springer Fachmedien Wiesbaden GmbH, part of Springer Nature 2019
V. Köpsel, *New Spaces for Climate Change*, RaumFragen: Stadt –
Region – Landschaft, https://doi.org/10.1007/978-3-658-23313-6_5

connected with its maritime location, its history of human settlement, as well as the high visual qualities of its landscapes. Cornwall is considered a beautiful coastal region with "undulating" hills (I-4: 2), scenic estuaries (e.g. I-2: 2, I-11: 2), a rugged coastline (e.g. I-5: 1) (see Figure 10) as well as important cultural and industrial heritage from past eras. All interviewees agree that the area's landscapes are visually "unique" (I-2: 1), "stunning" (I-5: 2), "extraordinary" (I-7: 2), and very different from the rest of the UK (I-17: 3). A central characteristic of these landscapes is perceived to be their variedness despite the relatively small land area of Cornwall (I-2, I-6, I-11, I-13). Cornwall's long coastline, surrounding the peninsula on three sides, thereby constitutes a defining element of both the landscape and regional identity, exposing it to the sea and the weather.

Figure 10: The rocky shapes of the south coast at Praa Sands, Cornwall
Source: own photo (2015)

Its exposed geographical location as a peninsula is mirrored in the interviewees' descriptions of Cornwall as "wind-swept" (I-19: 2) and in "battle between the landscape and the weather" (I-5: 2). These characteristics are not only visual features of the landscape, but are also very closely connected with the way in which the interviewees physically experience Cornwall. As I-7 (CoaST) describes, the Cornish landscapes have

> "a rawness and a realness [...] that you can't find in lots of other places. [...] It's uncomfortable and cold and hot and wet and windy and edgy. It's really, really edgy – both the landscape and the people." (I-7: 3)

Whereas the Atlantic coast in the north is described as "dramatic and a little bit more edged and wild" (I-18: 4), the south coast is associated with being "quite a lot calmer" (I-18: 4) and "sheltered" (I-11: 2). Other important landscape features in Cornwall are the moors. Especially Bodmin Moor, the largest of the moor plateaus in the area, is associated with isolation, solitude and a visual openness of the landscape (I-2, I-6, I-13, I-17, I-18). I-18 (AONB) appreciates that "if you want to get away from everything, you can go for a walk on Bodmin Moor and not see a single person all day" (I-18: 4).

When it comes to the human-made facets of Cornwall's landscapes, the central elements are perceived to be farmland and all landscape features related to it, as well as the relics from the mining era. Described as typical for Cornwall's farming landscape are small fields separated by Cornish hedges, probably one of the most prominent landscape features in the area. These hedges are found to be "culturally and environmentally important to Cornwall" both by the Council and by a number of interviewees as they provide shelter for local wildlife and reflect the long-standing history of human settlement and land management in the area (Cornwall Council 2017d) (I-3, I-4, I-8, I-10, I-11, I-18).

Figure 11: Ruin of an engine house, north coast of Cornwall
Source: Free Images Stock Photos (2018)

Other human-made elements commonly referred to are the relics of the former mining era: old engine houses and deserted mine shafts still visible in the landscape (e.g. I-2, I-6, I-17) (see Figure 11). Past activities of tin and copper mining are an important part of what still constitutes Cornwall's regional identity today (cf. 1.3.1). The mining heritage is marketed as a central component of the local landscapes by tourism organisations and Cornwall Council, reinforced by its regained fame through the TV series of 'Poldark' (see also 5.2). I-2 (CCFF) also assigns high importance to this facet of the area's landscapes:

> "If you turned around to somebody and said 'Can you draw a picture of something what identifies Cornwall to you?', chances are that they'll draw a picture of a mine!" (I-2: 14)

In addition to these physical features of the landscape, there are a number of shared associations made with the geographical location of Cornwall. A notion thematised in many of the interviews is that of being "at the end of the land" and a "feeling of isolation" from the rest of the country, especially from London and the UK's central government (I-18: 3) (also I-1, I-2, I-5, I-6, I-7, I-12, I-13). I-1 (NE) experiences this isolation positively and perceives Cornwall as a part of England "with its own language and its own cultural identity" (I-1: 1), whereas others emphasize the remoteness from the political "power base of the country" (I-12: 5). I-16 (NT), moreover, confirms the importance of being 'at the end' for Cornish regional identity in general:

> "It's that kind of 'end of the line' mentality. It's the Tamar River – then you're in. It's almost an island, but not. [...] If you're coming from up country, you have to go through lots of places and the last place you get to is Cornwall. So I think there's a whole historical kind of sense of its own country, so to say." (I-16: 3)

It is not only in terms of its geographical location that the landscape plays an important role in Cornish identity. The lifestyle in the region is also very much influenced by spending time outside conducting activities such as being by the sea or going for walks. The landscape is an important component of people's everyday lives and leisure activities (I-16: 16). Emphasising also the landscape's economic significance, I-12 (EA) argues that

> "in one way it's everything, isn't it? Particularly the coast I think. Gives the region its primary brand if you like – the coast, the beaches, the cliffs, I think, the exposure and the remoteness." (I-12: 9)

Especially the area's tourism industry relies on these qualities for attracting millions of visitors to Cornwall each year (I-4, I-6, I-7) (cf. sections 1.3, 5.2). For I-4 (Cornwall Councillor), the local landscapes are thus "very much a selling point of Cornwall as a tourist destination", especially during the summer holidays, for hikers, and for surfers (I-4: 6).

5.1.2   Shared Perceptions of Landscape Change

When it comes to changes perceived in Cornwall's landscapes throughout the past decades, there are several topics that are consistently named by the interviewees. Most commonly mentioned is the fast increase of built and housing development that both enlarges existing settlements and builds on previously untilled land (e.g. I-3, I-4, I-7, I-10, I-15). This rapid increase is perceived negatively, especially where formerly green spaces are sealed with concrete. I-10 (Par'lor) finds clear words for this development. In her eyes, "Cornwall has changed dramatically over the past thirty years and it's detrimentally bad on the landscape" (I-10: 6). With this increase in built development comes an expansion of the road infrastructure which is also perceived negatively by many interviewees (e.g. I-4, I-6, I-7).

Another form of landscape change in Cornwall is the increase in renewable energy structures in the county. As the peninsula is among the most windy and sunny places in England and profits from the UK's feed-in tariffs for wind and solar energy, it is very attractive for renewable energy developers (Cornwall Council 2013b). Having significant visual impacts on the landscape on the one hand and providing green energy on the other, the topic of renewables is contested among the local population (I-3, I-8), and inseparable from the debate about climate change. It is a topic addressed by almost all study participants (e.g. I-2, I-3, I-6, I-8, I-13, I-15); however, the perception of these renewable energy structures varies widely between the interviewees.

A third topic commonly brought up are changes in the local farming sector (I-6, I-12, I-13, I-15, I-17). In recent years, more and more fields have been joined together through the removal of the locally-typical Cornish hedges for larger-scale and mono-cultural cultivation of potatoes, daffodils and other flower bulbs (e.g. I-13: 6). This development is seen critically as it visually changes the farmed landscapes of Cornwall, but also because wide areas of Cornish farmland are cropped by corporations from outside the area to grow plants which are then, too, exported to outside of Cornwall (I-9: 7).

As became clear throughout the past paragraphs, the constructions of the local actors interviewed during the first empirical phase show a high number of commonalities. However, looking under the surface of these corresponding descriptions a number of very different, sometimes contrasting narratives unfold about what the Cornish landscapes are, what changes they undergo, and how they should be managed in times of a changing climate. The following sections present these different narratives in more detail, beginning with the ways in which Cornwall Council, the county's government, portrays the Cornish landscapes. I thereby show how differently Cornwall's landscapes are interpreted between different actor groups, and how these varying interpretations constitute a challenge for finding a joint approach to climate change adaptation.

5.2 The Official Policy Narrative

The narrative about Cornwall's landscapes officially promoted by Cornwall Council[13] can be viewed in its policy and strategy documents and demonstrates

---

13   For practical reasons, the policy narrative focuses on the perspective of Cornwall Council, the county's unitary authority, only. I do acknowledge that organizations such as the National Trust and Natural England are also shaping the wider policy context in Cornwall. As their

the significant role that the topic of landscape plays in Cornwall's regional identity and economic development. This narrative focuses on distinctiveness and diversity, natural and cultural heritage, and an ethos of preservation. Cornwall's landscapes are presented as shaped by human activities based on the area's natural resources over thousands of years. On the Council's website, the landscape is described as "the relationship between people and place. [...] It can mean a patch of local green space as much as a mountain range. The [...] landscape is stunning, diverse, unique" (Cornwall Council 2016c). Using the slogan "Think Cornwall, think Landscape!" (ibid), the Council assigns the local landscapes an important role by equating it with the entirety of what constitutes Cornwall. From this perspective, the Cornish landscapes serve a threefold purpose: they are an important element of regional identity, they shelter wildlife and cultural heritage, and they underpin Cornwall's economic activities. The recently drafted 'Cornwall Local Plan 2010-2030', moreover, clearly connects the visual attractiveness of the Cornish landscapes with the potential economic profit they provide by "attract[ing] locals, visitors and businesses" (Cornwall Council 2016b: 9).

A central strategy for creating higher economic revenue from the landscape is the preservation of local distinctiveness and characteristic building styles and the protection of cultural heritage sites. The Council's portrait of Cornwall's landscapes thus holds a clear policy imperative (ibid). Furthermore Cornwall's recently published Environmental Growth Strategy states as its primary goal that "[i]n 2065, Cornwall's environment will be naturally diverse, beautiful and healthy, supporting a thriving society, prosperous economy and abundance of wildlife' (Cornwall Council 2017b: 2). With this phrase, the document moves the Cornwall Council's interpretation of the local landscapes to the very centre of policy- and decision-making in the spheres of environmental management, planning and economic activity. This policy relevance is underlined in the introduction of the Environmental Growth Strategy which states that "Cornwall's brand is defined by our location and unique environment; it has provided the resources and provocation for innovation that made Cornwall globally renowned" (ibid: 4). While presenting Cornwall's landscapes as a visually attractive mosaic of natural and cultural heritage, the official policy narrative factors out past and present exploitation of natural resources and environmental damage through, for instance, former industrial activities. Instead, the unpleasant facets of the region's past are covered with a romanticized construction of a people living and working in harmony with beautiful landscapes. Regarding the management of these landscapes, this perspective entails a clear imperative to protect

---

understandings of the Cornish landscapes differ from those of Cornwall Council, however, I chose to represent their views through the narratives that I will contrast to the official policy narrative in the following subchapters.

their historic character against inappropriate development and greying-out of local distinctiveness (Cornwall Council 2011a, 2014a, 2015a). In this new document, however, the focus of economic activities in the area for the first time lies on sustainable growth and maintaining a healthy natural environment (see e.g. Cornwall Council 2017b: 30). The Council's focus on preserving the attractiveness of Cornwall's landscapes while fostering economic growth comes with an ambivalent relationship to change. Whereas on the on hand the importance of visually appealing landscapes as an economic resource is acknowledged, a significant number of locally contested large-scale housing and renewable energy projects has been permitted by the Council's planning department in recent years (Cornwall Council 2011a). Torn between preservation and economic development, the local landscapes are seen as both the basis of Cornwall's regional identity and as well as its main economic resource. Therefore, the policy narrative provides indications of unresolved tensions and suggests the co-existence of a plurality of landscape constructions informing the policy approach in the area.

Climate Change and its Impacts on Cornwall
As the portrayal of landscape change and priorities for management in the policy narrative is ambiguous, so is its perspective on climate change. Listing it as the main driver of landscape change in Cornwall, the local Environmental Growth Strategy assigns particular significance to the changing climate (Cornwall Council 2017b: 4). The Council, furthermore, underlines the "need to protect the quality and natural beauty, including the landscape [...] for its own sake but also as an economic driver and to build and maintain resilience to climate change" (ibid: 17). Although climate change is viewed as a potential threat both to Cornwall's heritage and settlements (Environment Agency 2012b, Cornwall Council 2015b), references to climate change in the Council's policy documents appear to be superficial in terms of actual practical recommendations. Also the Environmental Growth Strategy does not go into detail about the topic aside from stating that "[e]nvironmental growth will [be] the best opportunity to adapt to climate change, minimising harmful impacts and maximising the opportunities" (Cornwall Council 2017b: 35). Coordinated and joined-up efforts to climate approach adaptation by the Cornish government thus appear to be in their early stages and as yet an official or transformative perspective on the topic has not been developed in Cornwall (cf. section 1.3.2).

Whereas a consistent perspective on climate change with concrete implications for landscape management cannot be found in the policy narrative of the Council, four other narratives about the region's landscapes co-exist with that by the local authorities. These co-existing narratives provide important insights into different approaches to landscape management with direct relevance for adaption to climate change. To be able to highlight these implications for adaptation, I now present the four narratives by detailing their perspectives on landscape change and their imperatives for landscape management.

5.3 Co-Existing Landscape Narratives in Cornwall

The four narratives identified from the empirical data conceptualize the Cornish landscapes as human-environment interaction, natural systems, visual beauty, and functional production spaces. The following section goes into detail on what the landscape is understood to be in each narrative, what its constituting elements are, how the relationship between nature and society is understood, and what priorities for landscape management result therefrom. At the end of each section, I highlight the different understandings of climate change and its impacts. Tables for each narrative emphasise their key points. In addition, Appendix 8 gives an overview of the interviewees from both empirical phases and the different narratives identified from their responses.

5.3.1 The Human-Environment Interaction Narrative

In the human-environment interaction narrative (hereafter also HE narrative), Cornwall's landscapes are viewed as an interplay between human activity and the natural environment. The landscape is seen as a reflection of a long history of human settlement, a melding of natural and human-made elements, and a place where local communities live and work. It thus mirrors the ways in which Rose & Wylie (2006) conceptualize landscape as "embodied, perceived, affected" places of dwelling (ibid: 475). In line with this perspective, I-18 (AONB) describes the Cornish landscapes as

> "a baklava. [...] You start with geology and the plant cover, [...] and the views and the visibility; and then you've got the whole semi-natural habitats and the historic environment, and how the land has been changed through thousands of years." (I-18: 7; similarly I-7: 2)

When it comes to the human-made elements of the landscape, I-5 (VC) emphasises three types of heritage in Cornwall: the relics of the mining era such as old engine houses; small coastal villages reflecting the once strong fishing communities; and large buildings such as Truro Cathedral and old manors that are reminders of the past aristocracy in the area (I-5: 3) (see Figure 12). Clear connections between today's landscape and the impacts of human activity are drawn also by I-17 (EA). For him, particularly water in its different shapes is a central feature of Cornwall's landscape and has influenced human practice for centuries:

"We're surrounded by the sea on virtually all sides. [...] The expression 'Cornish sunshine' for when it rains, have you heard of it? [...] We've built our homes, our businesses, our cultures [...] around water. Whether it's around fishing, whether it's around access to rivers for moving materials – a lot of our lives' activities are associated with water." (I-17: 3)

Figure 12: Truro Cathedral
Source: own photo, 2016

I-13 (NT) refers to this interplay between natural and human-made elements of the landscape as "a lot of history of human habitation, so it's a very lived-in landscape" (I-13: 1). Also in publications by the National Trust and the AONB Partnership, the landscape is portrayed as a braid of natural and human-made elements showing the facilitation of an active use by locals and visitors as an integral part of the organisations' work (Cornwall AONB Partnership 2016b, National Trust 2016c) (see Figure 12). Human activities in the landscape thus stand in focus of this narrative and are seen as an integral factor shaping the present and future development of the local landscapes.

The human and the natural system are perceived as intertwined, jointly constituting Cornwall's landscapes and (re)shaping one another. At the same time, and similar to the policy narrative, there is recognition of the importance of the local landscapes for Cornwall's economic growth (I-6: 5, I-18: 8). I-18 (AONB) points out that

"the protected landscapes make provision for sustainable communities that live in the area. And you can't put a ring around the landscape and say 'We can't do anything in here' – people live there! [...] But we're still in that sort of old track of 'Yeah, but we can't do this, because it's protected!'" (I-18: 7/8)

This perceived connectedness of natural and human factors constitutes a central notion of the HE narrative: the critique of the separation of the human from the natural system, and thus of the detachment of society from nature (cf. section 3.1). Hand in hand with this critique goes the perception that the majority of the local population, and especially those who have moved to Cornwall to retire from 'up country' or own second homes in the area, do not understand how the Cornish landscape and its different elements function and should be managed; a point of concern when it comes to preserving historic knowledge about Cornwall's landscapes and how its features work in tune with each other (I-7: 8, I-10: 3, I-17: 6).

Figure 13: Portrays of the Cornish landscapes by the National Trust
Sources: ©National Trust Images/Paul Harris /Steve Haywood/Andrew Butler

Perception of Change and Management Priorities
The understanding of landscape as a preliminary result of human dwelling in the HE narrative leads to a high acceptance of change both through societal as well as natural processes (e.g. I-2, I-6, I-7, I-17). I-17 (EA) describes his understanding of changes in the landscape:

> "Something that you come to learn is [...] the way the landscape lives and changes. [...] We changed as a society, changed our practices [...]. And one of the things we do is making people understand that we live in a changing environment." (I-17: 7)

Whereas in the Council's policy narrative the historic landscape is viewed as an important economic resource in need of protection, this imperative for preservation is seen more critically in the HE narrative. The relationship between the landscape and its inhabitants is seen as reciprocal, and continuous change is a natural consequence of human dwelling. Although viewed as important for Cornwall's regional identity and landscapes, even the mining heritage is romanticised less than in the policy narrative, and its negative impacts on the environment and people's health find acknowledgement (e.g. I-4: 4, I-5: 27, I-6: 3). Since from the HE perspective humans and their activities are seen as part of the landscape, the growing amount of people coming to Cornwall is also perceived as inducing landscape change. For I-13 (National Trust), humans are not only changing the landscape by re-shaping its physical features, but also through the way they increasingly use it for leisure activities such as walking, cycling, and boating. Thus, he argues, the tourism industry and their call to 'use' the landscape in Cornwall is an important driver of change (I-13: 4).

Since the relics of past human activity are viewed as an important part of Cornish identity, interviewees following the HE narrative express concern about the 'greying-out' of the distinctiveness of locally typical features through, for instance, large-scale housing and supermarket developments (cf. section 1.3.1). Other than in the policy narrative, however, this distinctiveness refers not only to the historic relics of past mining and fishing activities but sets emphasis also on current local practices (I-5: 2, I-7: 25, I-11: 4, I-18: 10). For I-18 (AONB),

> "[i]t's that sort of 'death by a thousand cuts' kind of story where it's just that creeping, creeping greying out of character. You see it in High Streets and you see it in the tourism destinations and markets. The old local surf shop has now become Quicksilver, and the little old café is now COSTA." (I-18: 10)

She perceives not only the built features of local settlements as parts of the landscape, but also practices like the operation of stores and gastronomy owned by members of the local communities. Likewise, I-7 (CoaST) mourns the *"dismantling and designing out* [of] *that distinctiveness"* (I-7: 24). They both feel very personally affected by this gradual overtaking of 'sameness', as I-11 (CWT) terms it, and expresses disappointment towards this development (I-7: 24, I-18: 12). Consequently, the primary aim of landscape management in the HE narrative is to embrace and work with change whilst preserving local distinctiveness through sustainable, locally embedded human practice in the landscape.

Climate Change and its Impacts on Cornwall in the HE Narrative
From the perspective of the HE narrative, climate change is accepted as an existing phenomenon and perceived as already affecting the Cornish peninsula. Physical climate change impacts on the area are seen to be more frequent and heavier rainfall, river and sea flooding, coastal erosion, and increased storminess. These impacts are understood primarily in terms of their effects on human settlements and activities, so far affecting local communities more strongly than being clearly visible in the natural landscape (see e.g. I-7: 20/21). For I-17 (EA), the impacts of climate change become apparent in "the nature of the place where we live. It's that connection between where people live, where people's businesses are, and water." (I-17: 4). Similarly, I-18 (AONB) remembers:

> "When we had the storms in 2014, they did some damage at the coast and historic harbours, heritage beaches. [...] Those little fishing communities had it badly, their slipways damaged. Lots and lots of coastal communities were actually really suffering, the fishermen not being able to go out because it was too stormy." (I-18: 15)

These 2013/14 winter storms and related flooding incidents are an event often referred to in connection with the impacts of a changing climate on Cornwall(cf. Met Office 2014). Among the different organisations involved in landscape management, climate change is seen as one of the current foundations of landscape and environmental management (I-6: 13). However, there is disagreement between different organisations in Cornwall when it comes to how exactly to adapt to the changing climate (I-7, I-16, I-18). When asked for current conflicts around climate adaptation in Cornwall, a number of interviewees highlight the example of Godrevy on the north coast Cornwall; the headland that serves as this dissertation's in-depth case study and where coastal erosion has proceeded quickly in recent years (cf. section 1.3.3). A number of the local organisations have differing opinions about how to respond to these changes of the coastline (I-7, I-16, I-18), and the Godrevy case is viewed locally as a first example of disagreement about climate change adaptation in Cornwall (see chapter 6).

The diversity of perspectives on the Cornish landscapes is also mirrored in the individual perceptions of climate change impacts by the different interviewees. Although all interviewees who follow the HE narrative state that Cornwall is already affected by climate change, *which* physical effects are perceived as predominant differs. I-6 (Counc) and I-2 (CCFF) believe that the largest impacts of climate change on Cornwall will become visible in only the future (I-6: 4; I-2: 6). I-13 (NT) perceives current and future impacts of climate change as occuring primarily along the Cornish coastline (I-13: 16). I-18 (AONB) clearly connects flooding and coastal erosion to climate change:

> "More floods, more rain! […] The last few years have been ridiculous. Roads turning into rivers before your eyes and staying like that for months […]. If we had any kind of heavy rain, the fields were so wet from the winter that you just had gullies appearing overnight. […] And it would normally take months for that to happen." (I-18: 13)

Both the topography of the landscape and the geographical location of Cornwall are seen as contributing to the intensity with which climate change is affecting the area (I-2, I-13, I-17, I-18). For I-2 (CCFF), the increase of flooding incidents is connected both to climate change and to the fact that so many Cornish settlements are built close to river banks (I-2: 2). The impacts of climate change, he argues, are reinforced by mismanagement of human-made structures in the landscape, for example if drainages and culverts are blocked due to low maintenance and budget cuts (I-2: 3) (cf. section 5.4.1). Referring to the lack of knowledge about the local landscape thematised in the previous section, I-10 (Par'lor) underlines that ignorance about certain landscape features can cause severe problems in the context of river flooding:

"Stupid people moved into one of the other houses who wouldn't speak to any of us
[…], and they won't lift the sluice gate and they've planted up behind it. So [the
flood water] won't go down the lead anymore." (I-10: 12)

Also the perceived personal affectedness by climate change differs among the
interviewees. While I-6 has not noticed any changes to his day-to-day activities
because of the changing climate (I-6: 8), I-7 (CoaST) speaks in particular detail
about changes in the landscape and the weather; showing that she reflects very
critically on her experiences of the changing climate:

"Things are changing all the time […]. I remember a couple years ago this incredibly
hot February day and we were sitting there in t-shirts […] when usually it was absolutely freezing cold. But you were engaging in it in a way that you wouldn't normally
do on a February day, because it was very warm and sunny." (I-7: 15)

Table 5:   Overview of the Human-Environment Interaction Narrative

Phenomena addressed	Emotional articulation	Perception of change	Climate change & impacts
Landscape as interplay of human practice and nature; Result of human dwelling; Place for communities to live and work in; Expression of past human activity and local distinctiveness; Significant cultural heritage of fishing and mining era	Attachment to natural and human-made elements; Landscape as part of identity; Dislike of 'greying out' of distinctiveness; Cornwall as a unique place due to past and present human activity and natural beauty	Landscape as lived-in space; Change is natural result of human dwelling; Critique of loss of local distinctiveness; Call for sustainable human practice in the landscape to manage change	Climate change impacting on human settlements and activities; climate change accelerated by natural landscape *and* human structures and practice; Lack of landscape knowledge is problem for handling the impacts of climate change
Organisations: National Trust, AONB, EA, Visit Cornwall, CoaST			
Human activity = Dwelling, activities shaping and being shaped by landscape			
Landscape Mgmt. = Embracing change, sustainably preserving distinctiveness			

Source: interview data

Summarizing the Human-Environment Interaction Narrative
In the HE narrative the Cornish landscapes are understood as an interplay between its natural and human-made elements, one influencing the shape of the
other (see Table 5). The landscape features referred to reach from natural ones
such as rivers, valleys or the coastline to built structures like houses and roads.

Whereas the preservation of locally distinct building styles and human practice in the landscape is found important in this narrative, the main focus of landscape management lies on embracing the changes that are natural byproducts of human activity in Cornwall and responding to them through sustainable human practice. The interviewees following this narrative express attachment to both the natural and human-made elements of the landscape whilst being critical of an overly romanticizing perspective on the landscape. Moreover, they all perceive the combination of Cornwall's geographical location, its topography *and* human practice as accelerating the impacts that climate change has on the area.

5.3.2   The Natural Systems Narrative

> *"The landscape is [...] like a Rubik's Cube,*
> *and we're lost in one-side-of-Rubik's Cube-fixing."*
> *(I-7, CoaST: 26)*

In the policy narrative by Cornwall Council (cf. section 5.2), Cornwall's landscape is seen as a representation of natural beauty and past human activity on the one hand and an economic resource on the other. The natural systems narrative (hereafter also NS narrative), however, conceptualizes Cornwall's landscapes in terms of ecosystems and wildlife habitats, fields, wetlands, and a distinct assemblage of plants and animals. Referring to ecosystems types with clear boundaries, this narrative represents a classical natural-scientific, positivist understanding of landscape (cf. section 3.1). The Cornish landscapes are viewed as a sensitive natural environment under threat and in need of protection from negative human impacts. The wording circles around ecosystems and biodiversity, heathland and semi-natural habitats, habitat connectivity, rare species, and the protection and conservation of wildlife (I-1, I-7, I-10, I-11, I-12, I-15). To I-11 (CWT)

> "the way the landscape looks is largely a result of the wildlife and of land use. Many of the areas that you think of as being important for the landscape are also really important from a wildlife and biodiversity point of view; coastal habitats, moorlands, woodlands." (I-11: 3)

Thus the Cornish landscapes are described as "a mosaic of different habitats that are very close together" (I-10: 1). The prominent mining heritage of Cornwall is also mentioned in this narrative; however, not with a focus on local distinctiveness and its historic significance for Cornish identity. In contrast, the old mines with their shafts and residues of minerals are recognised in terms of the problems they cause in relation to flood management and pollutant washout, and with regards to the unique, barren habitats that developed on past mining spoils (I-9: 2,

I-11: 6, I-15: 1/2). The way in which the landscapes of Cornwall are constructed in the natural systems narrative is mirrored in publications by Natural England and the Cornwall Wildlife Trust.

Figure 14: Portraits of the Cornish landscapes by Natural England
Source: Natural England 2009: 1, 9, 13

A striking commonality of all these publications is that they present Cornwall's landscapes by using motifs of the natural environment and its wildlife almost always free of built infrastructure or human settlements (see e.g. Natural England 2009, 2011, Cornwall Wildlife Trust 2016) (see Figure 14). This separation between the natural and the human is symptomatic for this narrative. The natural and the human system are perceived as clearly distinct, and what is understood here as landscape are the natural elements of Cornwall's physical environment. This separation of the natural and the human comes with a rhetoric of protection of and custodianship over the landscape and an emphasis on the health of local ecosystems. In this context, I-7 (CoaST) explicitly articulates her thinking in terms of different systems that, taken together, make up the landscape. Throughout the interview, she underlines the importance of "understanding the whole system" (I-7: 1), refers to the "natural system" (I-7: 24), the "soil system" (I-7: 8), the "weather system" (I-7: 10), speaks of climate change as "a complex system" (I-7: 16) , and criticizes the overall lack of "systems thinking" (I-7: 12) within local government.

The interviewees following the NS narrative express attachments primarily with natural landscape elements (e.g. I-1: 1). I-15 (UT), for instance, states that he would be sad if certain local habitats would vanish with a changing climate (I-15: 5). I-10 (Par'lor), on the other hand, feels attached to the landscape through her hobby of bee keeping (I-10: 2/3). This focus on wildlife and ecosystems is mirrored by the good knowledge of the interviewees of local animal and plant species and the conditions they thrive in, as well as about how different elements of the natural system work in connection with each other (I-1: 1, I-7: 8, I-15: 9).

Perception of Change and Management Priorities

With its strong focus on preservation and sustainability, nature in the NS narrative is perceived as a network of fragile ecosystems in need of protection from harmful human activities. Chang is viewed as a positive process leading to alterations in habitats as long as it is induced by natural causes. Negative impacts on ecosystems and wildlife habitats through unsustainable farming and mismanagement of the land, however, are seen as the main dangers to a healthy landscape and referred as threat and intrusion (e.g. I-1: 3/4, I-15: 2). Likewise, built development and infrastructure are seen as impediments to an intact and sustainable landscape, and the rate of the increase of such structures in Cornwall causes concern (I-1: 4, I-10: 17, I-11: 4). As the vulnerabilities of Cornwall's natural environment are largely viewed as caused by human-made changes, it is understood to be the responsibility of management interventions to maintain and improve the resilience of wildlife and habitats. Thus, the central goals of landscape management are to create a resilient natural environment and to reduce adverse human impacts such as pollution and bad farming practice (e.g. I-1: 4). I-1 (NE) explains that "one of the reasons why we do what we do [is] to look after [the landscape]. We're looking after it. That's very important", thus expressing feelings of stewardship and responsibility for the landscape (I-1: 14). Similarly, I-10 (Par'lor) criticizes the Council's approach to planning, namely treating each planning application individually instead of considering the implications for the wider landscape, its sustainability, and distinctiveness. (I-10: 10). Interestingly, in the context of protecting the natural elements of the landscape, there are distinct ideas about what kind of nature is acceptable and what is not. This becomes especially apparent in how I-10 (Par'lor) criticizes the recent lack of management of habitats such as Bodmin Moor, Cornwall's largest moor plateau:

> "Places like Bodmin Moor [...] are not being grazed properly and things like that, so you're getting an encroachment of Gorsing and Racken. It really was quite desolate, and now it's just a semi-natural landscape. [...] It's going to end up being wooded again if they're not grazing it properly" (I-10: 6)

Certain types of plant species are viewed as undesirable for the Cornish landscape, and controlling them is an important imperative of landscape management. Thus, the preservation of local distinctiveness is also an central element in the NS narrative. However, this distinctiveness refers to the local flora and fauna instead of distinct buildings or cultural practice. Hence, the call for preservation in this narrative is limited to the protection of the natural environment which, as I-7 (CoaST) explains, is the basis for all human activity. She draws on a striking metaphor when criticizing the lack of joint-up and sustainable landscape management in the region:

"People, when they think of landscape, they just think of something visual. And not of something they live on and live in. [...] It's giving them clean air and clean water and something to eat. [...] For me, everything is like a Rubik's Cube! And people are like 'Oh, I must get all my yellow squares in order!'. But on the other side it's completely knackered. [...] And we're lost in one-side-of-Rubik's-Cube-fixing at the moment." (I-7: 26)

Climate Change and its Impacts on Cornwall in the NS Narrative
In the NS narrative climate change, whether it might be anthropogenically accelerated or not, is understood as part of a natural cycle and therefore seen as inevitable. Climate change impacts are perceived in terms of effects on the natural parts of the landscape such as wildlife and habitats, local species and the functioning of different landscape elements in tune with each other. An important topic in this narrative is therefore the encroachment of invasive species, for example different flowers, birds or jellyfish, that are draw to Cornwall by increasing air and water temperatures (I-7: 17, I-10: 10, I-11: 11). In addition to the migration of species, climate change impacts described in this narrative are more frequent heavy rainfall events (e.g. I-7: 17), greater stress for habitats due to changes in water availability or temperature (e.g. I-15: 9), and increased river flooding leading to the washout of pollutants from farms and mine shafts (I-1: 5, I-9: 7, I-15: 4). These impacts of a changing climate are being connected to the geographical location of Cornwall and its physical shape, as I-10 (Par'lor) explains. For her, climate change

"will make a massive difference because [...] we're right next to the sea and the estuary. So of course there is some coastal squeeze[14] going up and we're going to have more saltwater intrusion and it's going to push everything up [the estuaries]." (I-10: 8)

On the experiential level, the impacts of climate change on Cornwall described from the perspective of the natural systems narrative differ among the interviewees regarding their perceived personal affectedness as well as the spatial scale on which they are observed. I-11 (CWT) and I-15 (UT) refer repeatedly to what they term the 'landscape scale' (I-11: 10, I-15: 4); a spatial scale which, according to I-11, it is "greater than an individual site level" and is similarly conceptualized in publications with a natural-scientific outlook on landscape such as Natural England (2011). Although they do perceive the climate to be changing, for them the impacts are not yet noticeable in their day-to-day lives. Much rather, they argue, climate change is a long-term process that can be documented by

---

14   With the term coastal squeeze, she refers to what PONTEE (2013) defines as the reduction of space for wildlife habitat between insurmountable structures along the coastline such as hard-engineered sea defenses or steep cliffs one the one side, and a rising sea level on the other.

statistical data, but not experienced on a shorter term basis (I-11: 7, I-15: 7). Other interviewees, who have a more micro-scale understanding of the landscape, perceive how it changes on the basis of very small-scale alterations in their immediate surroundings. I-7 (CoaST) and I-10 (Par'lor) experience the changing climate based on observations they made in their own gardens, grasping climate change through the ways in which the flora and fauna changes, and thus display both their experiential and professional knowledge about the local ecosystems:

> "We have one apple tree and last week it had a huge apple and at the other branch there were apple blossoms. [...]That's not right at all, you know! That should be happening in April and this should be happening in October." (I-7: 16)

> "When I was younger, the seasons were very different than they are now. And the growing period is different for plants. We've got a very long growing period in Cornwall, something like 267 days. But it has definitely changed." (I-10: 9)

As different as their personal experiences with the changing climate might be, one particularity is shared by all interviewees following the NS narrative. Whereas both the policy as well as the HE narrative acknowledge the contribution that human activity makes to the acceleration of climate change (see 5.2, 5.3.1), here the causes of the changing climate are *not* being directly linked to anthropogenic activities. I-15 (UT) is convinced that "climate change is a real thing", however, "whether it's human-made or not doesn't really matter" (I-15: 8). Likewise, I-10 (Par'lor) explains that "it is going to happen anyway. [...] What's happening is that maybe we're accelerating it, but it's happening anyway" (I-10: 10). This neglect of the question of an anthropogenic acceleration of climate change is mirrored in the almost fatalistic attitude of I-5 (VC) and I-7 (CoaST) towards change in general and climate change in particular:

> "I think it's why we're in the trouble we're in: we've tried to beat nature. But that won't work." (I-5: 29)

> "We manage the changes as far as possible and encourage the right things to happen. [...] It won't help if we stood there and panicked, because all the change is happening and it doesn't matter how much we do that, it's going to happen." (I-7: 1)

Although I-11 (CWT) is convinced that climate change will have impacts on Cornwall in terms of changes to its fauna and flora, he thinks definitive conclusions about alterations in local wildlife cannot yet be clearly linked to climate change:

> "You get all those reports about butterflies and birds [...] changing, but they change with the weather anyway and it's only going to be over a longer period of time that we'll see whether those trends are truly moving in that direction or not. It's a bit too soon for me." (I-11: 10)

Table 6:   Overview of the Natural Systems Narrative

Phenomena addressed	Emotional articulation	Perception of change	Climate change & impacts
Wildlife, habitats and ecosystems; Farming, fields, land use; Hedges, heathland and wetlands; Farmland versus biodiversity; Local plant and animal species; Management at the landscape scale	Attachment to natural elements; Responsibility for healthy habitats; Concern about un-sustainable practices; Concern about invasive species and species change	Change positive if caused naturally; Change in farming practice emphasised; Critique of increase in built develop-ment; Human-made change and human practices endanger habitats	Climate change is happening and inevitable; Question of human role not overly relevant; Cc impacts understood primarily in terms of effects of flora, fauna, flooding, runoff

Organisations: Natural England, Wildlife Trust, EA, Parish Councillors, CoaST

Human activity = Outside the landscape, influencing it externally

Landscape Mgmt. = Protection and restoration of habitats and healthy ecosystems

Source: interview data

Summarizing the Natural Systems Narrative

Landscape in the natural systems narrative is understood as a braid of wildlife habitats and ecosystems the environmental sustainability of which is under threat by human interventions such as built infrastructure or pollution from farming. It is described in terms of natural elements such as fauna and flora, local and invasive species, nutrient flows and runoff rates. Interviewees following this narrative express attachment to and feelings of custodianship over the natural features of the landscape such as distinct species or intact habitats. Change is welcomed if induced by natural developments, but perceived as a threat if driven by potentially harmful human interventions. The focus of management thus lies on maintaining a healthy natural environment and reconnecting wildlife habitats to create a sustainable network of ecosystems in Cornwall. With regard to climate change, the question of the role of human activity is not seen as central because changes in the climate are natural and inevitable, and its impacts are understood in terms of effects on the natural elements of the landscape such as local wildlife or flooding and runoff regimes in the area (see Table 6).

5.3.3   The Visual Beauty Narrative

The third narrative identified from the interviews in Cornwall is that of the land-scape as visual beauty. Although the notion of visual beauty resonates in most of the interviews due to the appreciation that all interviewees share for the visual qualities of the Cornish landscapes, it can also be understood as its own narrative with distinct implications both for landscape management as well as for adaptation to climate change (see section 5.4). The visual beauty narrative (hereafter also VB narrative) reflects a classical romantic understanding of landscape as an aesthetically appealing stretch of land associated with an untouched natural environment and seen mostly in separation from larger human-made structures and settlements (cf. chapter 3). While most interviewees address the visual beauty of the landscape as a side notion of their landscape constructions, I-8 (Counc'lor) follows solely this narrative; with particular implications for his view on landscape management. He represents the classical understanding of landscape as a subject for painters and a place of romanticisation. For him, one particular visual characteristic makes the Cornish landscapes especially attractive:

> "The key thing that makes Cornwall different in my view is the light. You could understand why painters went down to St Ives or the west of Cornwall, because on certain days it is quite unique. Whether it's the salt in the air or what, I don't know – is really fantastic." (I-8: 1)

Although he perceives of both natural and human-made elements as constituting the Cornish landscapes, he makes clear that only certain types of buildings, namely those from the romanticized mining era, are worthy of preservation (I-8: 3). This distinct view on which human-made features are aesthetically tolerable and which are not is shared by Inter-viewee#18 (AONB). Although primarily conceptualising the landscape in terms of human-environment interaction, she perceives the visual implications of human dwelling as ambivalent: telephone poles or overhead wires, she argues, are eyesores and *"muck up this beautiful landscape"* (I-18: 4). When considering the aesthetics of Cornwall's landscapes, I-7 (CoaST) also prefers it natural-looking and undisturbed by human activity:

> *"I remember when I first drove through Cornwall [...] and saw the clay areas. I looked at the landscape and thought 'WHAT happened HERE? It looked so human-made. [...] That was very stark, very industrialized, heavy-duty. And that was quite shocking."* (I-7: 5)

In the case of I-8 (Counc'lor), his solely aesthetic perspective on the landscape shows little knowledge of, or interest in, the functions of landscape elements or their origins. Outlining his preference for the visual attractiveness of Cornish hedges, a feature described as an important part of Cornish identity by others (cf. 5.1), he refers to these structures derogatively as *"hedgerows or what they are"* and underlines how pretty they look from a helicopter (I-8: 4). Also when asked about the activities he undertakes in the landscape in his leisure time, his relationship with the local landscapes shows to be that of a detached bystander; a perspective of 'outsideness' in the words of Wylie (2007: 5) (cf. section 3.4.1):

> "We have dogs, but we walk them around town. [...] Occasionally we do go [on trips], but essentially it is to other towns. [...] I go to Parish Councils once a week. Sometimes I go to other visits for various reasons, to consult on something. So I go and visit a rural area." (I-8: 5)

The visual beauty perspective on Cornwall's landscapes has another side to it, however. A number of interviewees see an important function in the natural beauty and visual attractiveness which exceeds merely visual qualities: much as in the debate about the importance of blue and green spaces for human health in urban planning[15], I-7 (CoaST) highlights that

> "The outside here is where people go to regenerate themselves. If you're having a rough time, a bad day at work, a life crisis, you can go for a walk on the cliffs, on the moors, or our somewhere. And I think if you asked a lot of people, I think all of us do that." (I-7: 6, similarly I-5: 3)

This therapeutic notion, which is tied strongly to the visual qualities of the landscape, is very different from the detachment that I-8 displays. Here certain elements of the landscape are viewed as having healing qualities by re-acquainting people with their natural environment. Thus, the landscape in terms of visual beauty is interpreted quite differently by the different interviewees. The general focus on its aesthetic qualities nonetheless comes with particular implication for the perception of change and for landscape management.

Perception of Change and Management Priorities
In terms of landscape changes during the past decades, I-8 (Coun'lor) mainly perceived visual alterations through built development such as the construction of wind turbines and solar panels. When addressing the expansion of the renewables and traffic infrastructure, his concern lies primarily with the visibility of these new structures and not, as in the natural systems narrative, on potential negative impacts on wildlife and habitats (I-8: 6). His perspective on how the landscape will change in the future, moreover, reflects once more his attachment to its visual elements: "I don't think it will change very much. [...] I wouldn't want it to change! Because I think it's a magnificent, beautiful place." (I-8: 14). Although mainly following the HE narrative, I-16 (NT) also perceives the

---

15   For a review of evidence about the benefits of blue and green spaces for human health, see WORLD HEALTH ORGANIZATION (2016)

recently built wind turbines as negatively impacting the wild and attractive looks of the landscape:

> "I appreciate that kind of wildness and unspoiled... Some of the built things or developments that go on, I can see what effect they've had. That's been quite dramatic in recent years with solar farms and turbines." (I-16: 2)

I-7 and I-10 observed changes particularly to what they refer to as the "night landscape" (I-7: 10). Housing development, large superstores and new street lights are perceived as changing the night landscape of Cornwall which in past decades, due to an almost entire absence of street lights in the area, used to be pitch black in most locations (I-1: 1, I-7: 10, I-10: 18).

Moreover, local distinctiveness is an important facet of the VB narrative. While understood in terms of local practice from the HE perspective and local wildlife in the NS narrative, here this distinctiveness refers to what things in the Cornish landscape look like (I-8, I-13, I-16, I-18). As I-13 (NT) explains, one of these visual characteristics is the choice of materials used for building houses and infrastructure. He illustrates this with the example of a small fishing village in the part of Cornwall he manages for the National Trust:

> "Cadgwith, that's a good example. The beautiful, old, thatched buildings. Steeply sloping sides and very, you know, really fantastically beautiful Cornish village. But just around there it's very built-up with not very attractive, more modern buildings." (I-13: 2) (see Figure 15)

Thus, from the VB perspective human-made, non-traditional structures in the landscape that disrupt its natural and historic look are seen critically. The interviewees following this narrative express attachment with the natural and wild 'feel' of the landscape. This emphasis on attractiveness has strong implications for landscape management. I-18 (AONB) explains that her organisation orients towards landscape assessments with prevalently visual foci: the Landscape and Visual Impacts Assessment (Inox Group and Henry Boot Developments Ltd 2014), the Cornwall Design Guide for Buildings (Cornwall Council 2013a), and the Cornwall Landscape Character Assessment (Cornwall Council 2015a) (I-18: 16/17). The National Trust, too, puts strong emphasis on avoiding visually impairing structures in the landscape and builds with materials matching local building traditions:

> "We talk about good design for buildings [...] or trying to be sensitive to the characteristics of individual places. We try and get the right color stone or the right kind of stone. Or is it appropriate to use timber or metal gates? All those kinds of things. The vernacular." (I-16: 3/4)

Figure 15: Traditional building style in Cadgwith, Cornwall
Source: own photo, 2016

This avoidance of visually disturbing structures in the landscape is mirrored also in his personal view, as he feels emotionally uncomfortable about large, human-made, modern structures in the landscape:

> "There is that big radar dish at RAF Porthreath[16] [...]. When I came here you could look down the North coast and it was pretty much undeveloped the whole way. And then almost within a week this huge great thing turned up. Every time I look at that landscape I see it and it still winds me up a bit" (I-16: 4)

Climate Change and its Impacts on Cornwall in the VB Narrative
Since most of the interviewees who emphasize the beauty of Cornwall's land-scapes primarily follow one of the other three narratives, a broad perspective on climate change cannot be connected with the VB narrative. Interestingly, how-ever, the only interviewee who follows solely the VB narrative – I-8 (Coun'lor) – is also the only one who is skeptical about if climate change will have any impacts at all on Cornwall's landscapes at all (I-8: 7). He finds that

> "Obviously there have been big storms, so we've lost a few trees. We had some floods a couple years ago and that did change some of the land, damaged some of the properties. But by and large that has been addressed." (I-8: 6)

Although acknowledging that storms and flooding do occur in Cornwall, he does not perceive them as problematic or connect them to climate change. When asked if the changing climate has already had impacts on the landscape, he negates:

> "I don't think so. It's still a green and pleasant land. Possibly greener! We've had more rain. [...] But during the past years rainfall has been relatively modest." (I-8: 9)

---

16 Royal Airforce Base Porthreath is located on the north coast of mid-Cornwall (ROYAL AIR FORCE 2017)

Here again, in comparison with other interviewees, he shows to have a rather detached and superficial perspective on the landscape and evaluates climate change as much less dramatic. This perception is reflected in the fact that he states to neither feel any impacts of climate change in his personal life, nor names climate change as a topic of importance in his work as a Councillor (I-8).

Table 7:   Overview of the Visual Beauty Narrative

Phenomena addressed	Emotional articulation	Perception of change	Climate change & impacts
Visual attractiveness of natural and built environment; Lighting conditions; Traditional and distinct building styles; Eyesores in the landscape, e.g. renewables infra-structure	Attachment to natural looks and wild 'feel' of the landscape; Dislike of unsuitable or un-distinct buildings and structures; Dislike of renewable energy infrastructure	Change is seen critically if not being managed in a way that is suits the aesthetic requirements for the landscape	Climate change is not viewed as having noticeable effects on Cornwall and its landscapes; Neither affecting personal nor professional life
Organisations: Cornwall Councillor (AONB, National Trust)			
Human activity = Outside the landscape, detached, enjoying its beauty			
Landscape Mgmt. = Protection from inappropriate built development			

Source: own interview data

Summarizing the Visual Beauty Narrative

In summary, the landscape is portrayed in the VB narrative in terms of the attractiveness of the natural environment and, if not avoidable, built features. To keep the landscape as natural-looking as possible, built structures should consist of local materials and use traditional ways of building and landscape management. This imperative for visual attractiveness comes with a dislike of modern, non-traditional, built development. In terms of landscape change, the amount of newly-built housing, infrastructure and renewable energy structures in the landscape is viewed very critically due to their industrial and technical appearance. Hence, the main aim of landscape management is maintaining local visual distinctiveness by, for example, undergrounding electricity cables or preventing build development that does not match these ideas. Regarding climate change, particularly I-8 displays little knowledge of detailed processes in the landscape and thus does not perceive the changing climate to have any impacts on Cornwall now or in the nearer future; a perspective that reflects his relation to the landscape as an outside observer delighted by the beauty of the scenery (see Table 7).

5.3.4   The Functional Landscape Narrative

The three narratives presented above all acknowledge the attractiveness of Cornwall's natural environment as an important part of what constitutes the area's landscapes, although to different extends. The functional landscape narrative (hereafter also FL narrative), however, reveals a divergent perspective on the natural elements of the landscape and stands in strong contrast to the visual beauty perspective in particular. I-3, a Parish Councillor born in Cornwall and running an organic farm, has a perception of the place where he lives that is quite contrasting to the notions of human-environment interaction and natural systems. He sees the landscapes of Cornwall as considerably shaped through intensified farming, and thus as a basis for agricultural production and economic activity. Having a very functional perspective, his construction is free of the romantic and aesthetic notions often associated with the Cornish landscape:

> "I would describe it as a highly developed post-industrial landscape. [...] There is very, very little what you might call 'natural' about our landscape. I would think there is hardly one square foot of the county which has not been very, very heavily modified by men." (I-3: 1)

Also I-4 (Cornwall Councillor) has a very sober view of the area's landscapes. Although neither having a background in farming nor being originally from Cornwall, he shares the perspective of I-3:

> "All landscapes are to a great extent human-made. [...] The overall majority has been affected by men's activities over centuries and over thousands of years, there are not many primitive landscapes. [...] The landscape that we all see from our windows and cars is actually a food factory which has been crafted by men with hedges and fields." (I-4: 1)

In contrast to the NS narrative, in which the landscape is understood in terms of natural and semi-natural habitats, I-3 perceives the only areas in Cornwall untouched by human activity to be "the vertical parts of the cliffs" (I-3: 1). For him, the main purpose of this landscape is a functional one "by extracting a portion of the energy of the sun that lands on it off in the form of goods" (I-3: 2). Thus, the relationship of the natural and the human system is here conceptualized as a strong dominance of the latter over the former. If this dominance is not carried out sustainably however, it is seen critically in this narrative (I-3, I-4).

Similar to the policy and the HE narrative, both human-made and natural elements are perceived as part of the landscape here; however, neither the relicts from the past mining era nor contemporary building styles are viewed as central to regional identity. Hence, the romantisation of the mining era by the tourism sector through the designation of the Mining World Heritage Site is a point of critique in the functional landscape narrative:

> "You know, the Victorian times were a foul environment with all those mines and poisonous fumes coming out everywhere... And yet, since it's been redundant for 30 or 40 years, now these landscapes are viewed as iconic." (I-4: 4)

Especially on the side of I-3, the feelings expressed towards the landscape are very neutral. Asked if there were any places in the landscape that he felt particularly attached to, he admits: "I would say no. I'm very comfortable in this landscape. But particularly attached to it? No. I wouldn't say so" (I-3: 1). Feelings he expresses towards the local landscapes, and especially the Landscape Character Assessment, have a very critical undertone:

> "I am a sort of landscape cynic, I guess. And along with those landscape assessments, you got the Areas of Outstanding Natural Beauty which make up about 30-40% of the land in Cornwall. [...] I see them as really being a middle class construction in order to keep the poor people out of their areas." (I-4: 2)

His critique mainly refers to the clearly socially constructed boundaries in those assessments and the designations between different 'kinds' of landscapes since, he argues, the people living and working locally do not draw the same boundaries in their everyday practices. He thus inherently rejects the idea of landscape as something aesthetic that can be clearly divided into areas of different visual characteristics.

Perception of Change and Management Priorities
As perceived by the interviewees following the FL narrative, the landscapes of Cornwall have not changed significantly during the past decades (I-3: 2, I-4: 7). Change in general, however, is perceived as a positive and natural development, and the strong focus in the area on preserving certain landscape features is criticized:

> "Some people get really obsessed with preventing change. But you know, I think change is a good thing. We need to. And sometimes things will be better, sometimes things will be worse. We can't stand still, and so the landscape will inevitably change." (I-4: 8)

In his eyes, holding on to practices and structures from the 19th century and the strong backwards-facing focus of Cornish identity are barriers for a sustainable development of Cornwall and its economy (I-4: 1). Landscape perceptions with a strong romanticised notion such as in the Landscape Character Assessment or the AONB, so the argument, are artificial constructions that deny the contemporary needs of Cornish communities instead of constructively contributing to green growth and sustainable land(scape) management (e.g. I-3: 11, I-4: 2). Whereas heavily criticised in other narratives, the increase in renewable energy structures in Cornwall is here not perceived as having a negative visual impact on the landscape (I-3: 10, I-4: 5). Instead, the hesitation to implement innovative alternative energy sources is a central point of critique. This unromanticised perspective on the landscape comes with a strong critique of the efforts by the National Trust and similar organisations to preserve local distinctiveness and limit built development in Cornwall. It is emphasized that decisions should be made on the basis of what is rationally necessary to address the pressures local communities and the local environment are facing (I-4: 18). The main management aim is therefore to make ideal use of Cornwall's landscapes for farming and food production, whereas the focus on preservation of the relicts of past times is seen as an impediment for sustainable development rather than having a positive influence.

Climate Change and its Impacts on Cornwall in the FL Narrative
From the viewpoint of the FL narrative, the impacts that climate change is likely to have on Cornwall's landscapes correspond to the perception of the dominance of the human system over the natural. I-3 thinks that as a consequence of climate change, there will be "small changes […] from natural means and quite big ones from human." (I-3: 14). Climate change is perceived as clearly having impacts on Cornwall's natural environment, and visions for the future are presented with a wording of threat and uncertainty (e.g. I-3: 5). I-4 states that "climate change is a real issue and I worry deeply about it" (I-4: 10). In his eyes, the major impacts will occur where sea level rise and increased wave action threaten coastal villages (I-4: 10). Although being convinced that the climate is changing and action is needed he, personally, does not experience any effects yet:

> "I think […] the change in imperceptible for us really, isn't it? The daily fluctuations in temperature and rainfall would almost swamp the long-term effects. […] I am fairly prepared to accept that sort of evidence, but I honestly can't say that personally, I've noticed." (I-4: 15)

I-12 (EA), on the other hand, understands the impacts of climate change rather in terms of their effects on local farming practice and agricultural production, for example when heavy machinery cannot be used on soils anymore which are too saturated with rain water (I-12: 11). At the same time, there is also acknowledgement in this narrative that people's choices to settle on flood plains and near river banks accelerate the impacts of climate change to a similar extent as the topography and geographical location of Cornwall (I-4: 10, 22). I-3 paints a particularly dark picture of the future of the protected Cornish landscapes along the eroding coastline in the case that mitigation action on climate change is not taken seriously. The recommendation he gave to the Area of Outstanding Natural Beauty Partnership mirrors his cynical perspective on the romanticization of the landscape:

"I recommend to the AONB if they're not prepared to embrace carbon cutting, they need to invest in glass bottom boats so people can still see the outstanding landscape through the bottom of the boat." (I-3: 4)

His statement reflects the strong critique in this narrative of a focus on preserving the visual qualities of the landscape in times when a changing climate calls for action that overlooks romantic notions for the sake of adapting communities and farming practice to new environmental conditions.

Table 8:   Overview of the Functional Landscape Narrative

Phenomena addressed	Emotional articulation	Perception of change	Climate change & impacts
Farming, food production; Human-made landscape, industrialized; No pristine nature left; Landscape is subject to exploitation through mining and agriculture; Purpose of landscape is extraction of food and energy	Utilitarian understanding of landscape; No strong attachment expressed; Free of romantic notions; Disappointment in agencies for unsustainable management; Critique of backwards-facing regional identity	Change positive and natural; Engineered structures in landscape not visually impairing; Support of renewable energy; Strong critique of preservation of relics of the past	Climate change is real and threatening; Will become more serious in the future and impact on communities and farming; Cynical view on protection of landscape aesthetics when determined action is needed
Organisations = Cornwall Councillor, Parish Councillor			
Hhuman activity = Living off the landscape, replacing nature with economic activity			
Landscape Management = Providing food, energy and goods for people			

Source: own interview data

Summarizing the Functional Landscape Narrative
The landscapes of Cornwall are perceived as spaces of agricultural production and economic growth in the FL narrative. Lacking notions of nostalgia or romanticism, this perspective acknowledges the strong dominance of human activity over the area's natural environment and views human-made interventions in the landscape as justifiable as long as they are compatible with environmental sustainability. Aesthetic impacts are not given much consideration. The interviewees following this narrative do not express high attachment to the landscape in general or its aesthetic qualities in particular, and topics such as the preservative stance on landscape management by Cornwall Council and the AONB Partnership are met with a wording of cynicism and irony. Landscape change through both natural and human-made developments is widely accepted as a consequence

of human dwelling and natural forces, and strong criticism is voiced regarding the backwards-facing focus of Cornish regional identity. In this narrative, climate change is perceived of as a real and threatening process the effects of which will become more serious in the future and must be taken very seriously in any considerations around land and landscape management (see Table 8).

5.3.5 Individual Ways of Relating to the Landscape

Although interviewed primarily in their professional roles, all interviewees show to have individual and intimate relationships with Cornwall's landscapes. These relationships –the emotional dimension of individual landscape construction (Kühne 2018) – on the one hand demonstrate the significance that the phenomenon of landscape is given by the interviews also on a personal level. On the other hand, it became apparent that there are a variety of ways in which the interviewees personally relate to the landscape. As it becomes clear in the policy narrative (see section 5.2) and the other narratives presented above (sections 5.3.1 - 5.3.4), the local landscapes of Cornwall, however the details are interpreted, constitute an important facet of regional identity (e.g. I-2, I-6, I-7, I-15). In the eyes of I-15 (UT), these landscapes play "a big role […] in what people associate Cornwall with. The way it looks, the way it feels, the way it's green" (I-15: 3). The significance of his statement is reflected in the high attachment the interviewees express to Cornwall's landscapes generally and to specific elements of it (e.g. I-2, I-5, I-6, I-15, I-18). I-7 (CoaST) explains that in her view, the landscapes have intangible qualities that make people, even if not originally from Cornwall, feel a strong sense of belonging:

> "It's like a jewellery box that is ridiculously beautiful and full of places which are different. […] And […] that's your origin, where you're from and where you'll go back to. Even if you're not 'ethnically Cornish'. I'm not from here, but I am of here." (I-7: 4)

I-2 (CCFF) was born in Cornwall and lived away for 20 years before returning to his home county. For him, too, the area's landscapes are an integral part of what he feels attached to and "very much a subconscious part of why I love Cornwall. I think it's best described in my desire to come back" (I-2: 1). Four particular ways of how the interviewees personally experience the landscape became apparent from this study:

1. Individual/visual landscape preferences;
2. Childhood memories;
3. Leisure time activities;
4. Experiencing the landscape through the weather.

Among those interviewees who grew up in Cornwall, their childhood memories constitute a second dimension that strongly influences the attachments to the local landscapes (I-7, I-10, I-11, I-15). I-15 (UT) recalls spending a large part of his youth on the local beaches to which he still feels particularly attached (I-15: 3). For I-11 (CWT), it is the familiarity with the landscapes of Cornwall, where he was born and raised, that makes him feel at home in the area (I-11: 2). I-10 (Par'lor) feels strong emotional bonds to the landscape especially through memories of the places where she played with her siblings as a child (I-10: 2). Although not being from Cornwall originally, also I-7 (CoaST) appreciates its landscapes for reminding her of her home region in north Scotland.

A third way of relating personally to Cornwall's landscapes is based on the present rather than the past: the hobbies and activities that the interviewees carry out in their leisure time. Whereas I-11 (CWT) describes himself as having "always been a keen walker" who enjoys spending time outside, I-10 (Par'lor) and I-16 (NT) have formed a relationship with the local landscapes through their passions for wildlife by being a bee keeper and a bird watcher.

Remembering watching the waves on stormy days, I-18 (AONB) touches on the forth way of experiencing the landscape that becomes apparent from the interviews in Cornwall: the experience of the landscape through the weather. Experiences of bad and rough weather are seen as an expression of the exposedness of the Cornish landscapes and associated with a feeling of being alive and connected with the elements (I-2, I-5, I-7). I-7 (CoaST) connects her experiences with the "really unpredictable" weather with the sense of being "close to the edge" in Cornwall (I-7: 3). The strongest appreciation of the landscape and the weather, however, is expressed by I-5 (VC). Associating stormy weather with feeling the forces of nature and the elements, his landscape experience is coloured by sensual and acoustic:

> "It's quite rugged and robust and tough. You DO feel nature here! And I like it! I'm looking forward to wild nights walking on the beach in the dark with a torch and my dog in the winter [...] You might be in the dark, but you hear the sea and you hear the sounds of Cornwall. And you've got that roar when it comes in and it's really stormy. I love it!" (I-5: 1)

Reflecting the interviewees' intimate relationships with the landscapes of Cornwall, these four dimensions of personal experiences among professionals in landscape management are particularly interesting against the background of challenging the role and objectivity of experts (cf. section 2.3) and will be discussed further in chapter 7. In combination with the five narratives presented above, these different emotional dimensions of landscape construction strongly underline the importance of the topic of landscape both for the area's regional identity and for policy-making and planning in local government and organisations. While each story about the landscape is coloured with individual experiences, certain stories focus on very similar landscape elements and vocabulary and result in particular imperatives for landscape management. In the following, I expand on their implications for climate change adaptation before turning in the next chapter to the smaller-scale case study of Godrevy to set these narratives in the context of a particular place and its local debates.

5.4 The Narratives' Implications for Climate Change Adaptation

Each of the five narratives about Cornwall's landscapes outlined above comes with distinct implications for adaptation to the impacts of climate change. To set the broader context of these implications, I outline in the next paragraphs four factors that influence the wider context in which climate adaptation activities take place in Cornwall. Afterwards, I contrast the implications of the different narratives with each other and highlight the conflict potential for landscape management arising from these varying perspectives in times of a changing climate.

5.4.1 The Broader Context of Climate Adaptation in Cornwall

The interviews with Cornish actors in landscape management confirm the observation from the analysis of documents by Cornwall Council (see section 5.2): a joint, coordinated strategy for adaptation to climate change does not exist in Cornwall at present, neither in form of a publication nor in form of a position within the Council to coordinate existing adaptation efforts by the different organisations in the region (I-2, I-4, I-6, I-7, I-18). I-18 confirms the assumption that, although Cornwall's location as a peninsula reinforces the impacts of climate change, not much adaptive action is undertaken yet:

> "There's a more tangible sense of climate change happening here than elsewhere […] Is anyone doing anything to try and tackle it? Not much yet. I think that's where Cornwall is: that the recognition is there, but practical action needs to happen." (I-18: 19)

I-4, a Cornwall Councillor himself, sees a cause for this hesitant approach to climate adaptation in the conservatism and skepticism on the side of those elected to represent the different Parishes of Cornwall in the Council:

> "[I]n practice, we've done very, very little for [adaptation] and I suspect that the majority of elected Councillors are probably climate change skeptics. […] As an organisation, we haven't done nearly as much as we pretend we've done." (I-4: 11)

His perception of the hesitant attitude of his fellow Councillors mirrors a tension described as typical for Cornwall by a number of interviewees: the tension between "a culture that is really geared to innovative solutions" versus what I-3 (Par'lor) terms an "innate conservatism [...] to actually imagining a different way of doing things" (I-3: 12/13). Likewise, I-13 (NT) sees the reason for this resistance to change in the fact that

> "politically we're quite a conservative county. [...] We're not very radical, I suppose, so it doesn't feel like there's big changes politically that will shift things in a different way." (I-13: 16)

Although the past mining era has left the Cornish people with a spirit of innovation and expertise this conservatism, hand in hand with the strong urge to hold on to historic structures in the landscape, is seen by many interviewees as one of the main causes for the slow response to climate change in the region (I-3, I-4, I-6, I-7, I-13).

Another issue closely related to the inaction by the Council is perceived to be the lack of funding in the area (I-2, I-4, I-6). Working in the Environmental Planning Team of Cornwall Council, I-6 does not "think we've got an [...] own structure that specifically focuses on adaptation issues." One important reason for that are the budget constraints that increase every year under austerity politics. Upon the question why the post of Climate Change Officer does not exist anymore within the Council (cf. section 1.3.2), he explains: "It's budget-related. We'd probably still have that post if we didn't have to look after our priorities." (I-6: 12). These priorities, I-4 (Counc'lor) argues, currently do not lie with developing suitable responses to climate change as Cornwall does not "have the political leadership from the elected members who say 'This is a priority area'." (I-4: 13). As Cornwall used to receive large sums of money from the European Union for projects related to the environment and infrastructure, the fact that the UK is planning to leave the EU is likely to further increase these financial problems; probably resulting in more cutbacks in the Council's environmental and landscape management duties (The Guardian 2016). In contrast to Cornwall Council, however, a number of regional and local organisations such as the National Trust or Natural England have moved climate change adaptation high up on their agendas, as will become clear during the following sections.

A fourth factor influencing the efforts towards climate adaptation undertaken in Cornwall is grounded in its geographical position. As outlined in sections 1.3.1 and 5.1, an important facet of Cornwall's regional identity is its location 'at the edge' or 'at the end of the land'. This mentality, I-2 (CCFF) argues, results in a particular way of solving pressing issues in the area:

> "Cornwall is a long way from London. We're at the end here, so we have to fight for ourselves. And that has made us by nature very resilient and self-sufficient. So we're making stuff happen without that assumption that Government will sort it out for us.." (I-2: 13/14)

As will become clear in the following section, this attitude of independence and self-reliance has a clear influence on how adaptation to climate change is approached in Cornwall. Against the background of this wider societal context in which climate adaptation takes place in the region, I now return to the main points of the four narratives about the Cornish landscapes (cf. section 5.3) and highlight their distinct implications for adaptation approaches in the area.

5.4.2 Differing Perspectives on Climate Change Adaptation

As I outlined in chapter 5.3, the Cornwall case study shows that a common attachment to and high appreciation for a landscape do not necessarily result in consensus on how best to manage it under a changing climate. The landscape narratives clearly show that Cornwall as a place is interpreted very differently by the local actors; interpretations that have diverse implications for landscape management and for the question of what an adaptive response to climate change should look like in practical terms. Cornwall Council's Local Plan 2010-2030 underlines the "[…] need to protect the quality and natural beauty, including the landscape, […] for its own sake but also as an economic driver and to build and maintain resilience to climate change" (Cornwall Council 2016b: 17). However, a concrete operationalisation of what this protection or any other response to climate change could look like cannot be found in the official documents (see 5.2). The four landscape narratives derived from the interviews, on the other hand, have a variety of distinct implications for adaptation to climate change, which at times stand in surprising contrast with each other. In the following paragraphs, I present these different implications and highlight how they might constitute a root for misunderstanding and a barrier for the development of a joint adaptation strategy in Cornwall.

Implications of the Human-Environment Interaction Narrative
Landscape management in the HE narrative is seen as a process of co-adaptation of the landscape and human practice in it. Central to this narrative is a criticism of the detachment of society from nature and a call to reverse this separation (see 5.3.1). Regarding adaptation to climate change, this emphasis on lived-in landscapes results in a demand for locally-embedded, bottom-up initiatives and community projects that focus on the local population as key actors for change. I-17 (EA) sees an effective response to climate change in working closely with the affected communities that "will have a key role in adaptation because they understand their local landscape and also, importantly, they're there. They're at the place" (I-17: 1, similarly I-6: 12). Such a response to climate change can be achieved in forms of "community change and community acceptance and understanding, and then ultimately community-lead actions", he explains (I-17: 9).

Building resilience to climate change is thus perceived of as adapting the ways in which communities live in the landscape and shape the places where they live.

The fact that change is accepted as a natural by-product of human dwelling in the HE narrative is mirrored in the approach to climate adaptation resulting from it. For I-16 (NT), responding to climate change means not so much holding on to existing structures, but rather a necessity to embrace change and work with the occurring natural processes. In his words, the National Trust is managing landscapes "in that sense of 'Shifting Shores', as we call it, which is kind of working with natural processes" (I-16: 13) (cf. sections 1.3.2 and 6.1). Likewise, as I-2 (CCFF) points out, embracing landscape change means to let go of structures that are no longer feasible to maintain such as

> "Mullion Harbour[17] down in the Lizard, an ancient fishing port. [...] The National Trust is putting an awful lot of money rebuilding the harbour because of the effects of the weather on it. But [...] I think that is something a bit overprotected sometimes, that stuff. And we've got to better let things go." (I-2: 8)

I-16 (NT) speaks about adapting to coastal change in very similar terms. For him, adaptation is about "working with change, rather than saying it's like this and I want to keep it like this" (I-16: 12) because, as he explains, the forces of nature are out of human control in any case.

From the perspective of the HE narrative, a central component of working with communities around climate adaptation is to educate the local population about how different landscape elements interact with each other and about the causes and consequences of certain practices in the landscape. This attitude is reflected in the narrative's implications for climate adaptation in a focus on "re-engaging [...] the community with the place where they live" (I-17: 12). Conceptualising the landscape in terms of a common good, the responsibility for which is shared across society (cf. section 3.3), it is viewed as important to re-connect the local communities with the places where they live and hand over to them the task of managing them sensibly. This re-engagement is operationalised in terms of enhancing the little existing knowledge about the functioning of different landscape elements in tune with each other, causes and consequences of human practice in the landscape, and the historic background of certain features and environmental management practices. I-17 (EA) explains that

> "When we work with communities, we try to re-engage them with where they live. [...] We talk about that places where people live to [make them] understand flood risk [...] and understand the changes in the landscape through the seasons and through framing practice. And the impacts they can have as a community." (I-17: 6)

---

17   For a detailed account of the management of Mullion Harbour in times of a changing climate, see DESILVEY (2012)

Similarly, I-7 (CoaST) calls for capacity-building measures around current and future changes in the landscape both from natural and human causes in local communities. Her organisation works closely with affected businesses to prepare them through practical advice for incidents of flooding in their communities and provides information about possible protection measures against flooding (I-7: 19). Small-scale and bottom-up adaptation activities such as the re-orientation of floor levels in houses or the installation of flood gates are also demanded by I-6 (Counc) (I-6: 15). In the words of I-17 (EA), such measures need to be tailor-fitted to the specific locations where they are implemented as climate adaptation is "a bit like a jigsaw. There is not a 'one size fits all'–solution. Actually it's about lots of components, because often in a place there *is not* one [...] risk" (I-17: 12). I-16 (NT), moreover, explains that his overall perspective on landscape management and adaptation has changed in recent years:

> "Actually now I think we do need to be adaptable. [...] Rather than thinking 'Crystal ball, what's the solution I need in 30 years and build it now?' It's like 'No, there's probably about 3 or 4 different ways it could go and what options mean that we can sensibly choose those when we get closer to them'." (I-16: 9)

In that sense, landscape management under a changing climate means not only changing the way of shaping the landscape, itself, but also changing the entire management process towards a more adaptable approach. Although local distinctiveness is held to be important in the HE narrative, the overall perspective on adaptation is pragmatic and moves human dwelling in the landscape into focus. I-6 (Counc) underlines that for him, it is important that adaptation measures are locally fitted and provide a real benefit to the affected communities (I-6: 13). Restoring or maintaining older adaptation features such as historic flood alleviation schemes is important in this context (I-6: 11). Furthermore, I-7 (CoaST) suggests that ideally, measures such as flood gates should be planned and integrated in newly-built housing developments from the beginning on (I-7: 21).

Whereas other interviewees are skeptical about whether the strong regional identity and its rather conservative focus might be a constraint to climate adaptation in Cornwall (cf. section 5.1), I-6 (Counc'lor) points out the strengths of it for responding to climate change. Referring particularly to the 'end of the land'-mentality many people have (cf. section 1.3.1), he argues:

> "That's a typically Cornish thing: we're [...] at the end of the peninsula, and therefore there's a mind-set around being the last to receive the benefits and the first to lose the benefits. [...] There is that mentality that says building resilience into our communities is a really good way of approaching our climate change motivation. And I think trying to achieve community resilience is something that is being quite key here." (I-6: 14)

Also I-16 (NT) sees potential benefits of Cornish regional identity for adaptation to climate change, again emphasizing the importance of communities as partners in adaptation activities:

> "[I]f you could harness that kind of pride about Cornwall and it's rugged beauty and the nature of it, then perhaps that would allow you to get over some of the hurdles that you're always going to face anywhere. [...] People have got that pride in their area." (I-16: 15, similarly I-7: 24)

In summary, the adaptation approach in the HE narrative focuses on community engagement and small-scale adjustments in human practice. The goal is here to adapt both the landscape itself *and* the ways in which humans act in the landscape to the changing climate. Preserving local distinctiveness is perceived as important for Cornwall's economy and regional identity; however, landscape change as a product of human dwelling is widely accepted. The reconnection of the local population to the landscapes they live and work in, and to sustain local knowledge also about past solutions for issues such as flooding or coastal erosion, are viewed as very important.

Implications of the Natural Systems Narrative

> *"Climate change is a rip current. If we resist it, we're screwed."*
> *(I-5, VC: 28)*

From the NS perspective, responses to climate change should focus on creating healthy ecosystems through effective management of the landscape; a view shared by the Wildlife Trust and Natural England, but also other interviewees. Climate change is taken as a given in this narrative (see also 5.3.2) and, as I-11 (CWT) explains, it does not matter much if its causes are humanly accelerated or not. The imperative to manage the landscape sustainably goes hand in hand with a strong sense of stewardship over the natural environment. The NS narrative operationalises climate change adaptation by working with natural processes wherever possible and protecting the landscape from harmful human interventions such as engineered structures; especially regarding coastal protection and the alleviation of river flooding (I-10, I-11, I-15). I-11 (CWT) argues that working *with* nature instead of against it is the best alternative as several hard flood defences in Cornwall have recently been overtopped or breached by increased wave action. Climate adaptation to him is "letting nature reclaim some of those areas and to deal with the consequences" (I-11: 7). This demand to support natural solutions is mirrored in how I-5 (VC) puts in words his understanding of climate change referring to another phenomenon closely connected with the landscape, namely the surf culture of Cornwall:

"We have a saying among the surfers in Cornwall which is 'Never swim against the rip current, you'll drown'. To me, climate change is a rip current. If we resist it, we're screwed. If we learn to go with it and see where it takes us, there will be a changed landscape because nature will make it change – but [we are] not to resist it." (I-5: 28)

Natural solutions should be implemented in a bottom-up, small-scale manner. The spotlight here lies not on the local communities and their practices, but on working together with farmers, improving land management and reducing polluted runoff (I-11, I-15). Although working at the local level, these efforts are understood to improve the landscapes of Cornwall on a large scale.

An example of natural solutions undertaken that (re)create wildlife habitats is a project run by I-3 (Par'lor) who reintroduced beavers to Cornwall on a fenced pilot site on his farmland. With a river running across his land and a village downstream that is prone to flooding, this innovative way of restoring former flood plains is in tune with his focus on sustainable farming. In addition to environmental considerations, which were the main motivation for starting the project, the lack of funding from the authorities reinforced his decision to implement alternative low-cost flood alleviation (I-3: 7) (Cornwall Wildlife Trust 2018).

Besides this joint approach of cooperating with local farmers, another important way of making the natural landscape resilient to climate change is seen in (re)connecting wildlife habitats with each other to foster species migration (I-11, I-15). Concentrating more generally on building a healthy natural environment, I-11 (CWT) explains that his organisation understands climate adaptation as a process that increases the resilience of entire ecosystems so they can cope with the impacts that arise from the changing climate. For him, adaptation is yet another reason for improving environmental sustainability. The fact that climate change has received much attention in recent years is a weighty argument for building sustainability into all landscape management decisions (I-11: 8).

All in all, the approach to climate adaptation in the NS narrative is driven by an urge to carefully manage the natural landscapes and wildlife habitats of Cornwall, notions of responsibility for their environmental sustainability, and a sense of stewardship over the landscape. Engineered structures are rejected for reasons of maintaining a healthy local wildlife, and the primary focus of adaptive management lies on applying soft solutions to natural processes and reducing pollution from flood water and farming runoff.

Implications of the Visual Beauty Narrative
Aspects of the visual attractiveness of the landscape play a role in almost all landscape narratives outlined above, and the visual aesthetics of the landscape play an important part in the response to climate change in Cornwall in general.

This strong focus on aesthetics is mirrored also with regard to mitigation measures in how I-8 (Counc'lor) perceives the renewable energy infrastructure in the area. His attitude towards these structures shows solely aesthetic considerations without attention to their usefulness and also reflects his attitude towards engineered adaptation measures:

> "If you look down from a hilltop and you can see that farmland with the blue back-drop, you think 'How wonderful!'. And then you see the wind turbines which seem to be growing in size increasingly and the growth in solar panels, and people don't like that. They get irritated by that." (I-8: 2)

Whatever adaptation measures are carried out, for I-8 the most important characteristic is that they fit in well with the looks of the surrounding area. Similarly, the National Trust and the AONB Partnership both try to fit new adaptive measures to the current appearance of the landscape, as I-13 (NT) explains:

> "I think we'd still try and maybe make the balancing between natural beauty and adaptation measures. I think we're probably trying to find a way through, but whatever adaptation measure has been discussed, we'll be thinking about how it looks in terms of its natural beauty, and not too much use of concrete." (I-13: 12)

This strong emphasis on aesthetics results in a focus on adaptation measures as protecting "those wild landscapes and keep[ing] a kind of natural, traditional feel to them" (I-16: 13). Therefore, similarly to in the NS narrative, natural solutions are here viewed as central to climate adaptation,; however, the reasoning for this is very different from the NS narrative's emphasis on supporting ecosystems functions and not interfering with wildlife. As the visual beauty perspective is strongly featured also in the different management plans of organisations such as the National Trust or the AONB Partnership, it has significant weight in the context of climate change adaptation approaches in Cornwall. From the perspective of the AONB Partnership, this "distinctive character and natural beauty are so outstanding" that the "protected landscapes [...] are Cornwall's most important assets" (Cornwall AONB Partnership 2011a: 9), thus giving the preservation of their aesthetic characteristics a clear imperative also for policy and planning.

Implications of the Functional Landscape Narrative
Whereas the three narratives presented above center the natural beauty and local distinctiveness of Cornwall's landscapes, the functional landscape narrative has quite different implications for adaptation to climate change. Not connecting the landscape with romantic or aesthetic notions, the interviewees following this narrative do not perceive built features such as renewable energy infrastructure to have any negative impacts. On the contrary, I-4 (Coun'lor) sees the necessary response to climate change in engineered solutions:

> "In order to mitigate the effects of climate change we need engineering solutions. [...] And we do accept them in our day-to-day lives! We also accept road infrastructure which is a horrendous scar in the landscape."

Also I-3 (Par'lor) takes climate change and its impacts very seriously. Similar to the suggestion of I-16 (NT) in the HE narrative, he perceives future impacts of climate change to be severe yet uncertain. He thus attaches particular importance to adaptation approaches that are flexible and apply whichever measures are most effective without consideration of the iconicness of the landscape. His "fear is that we've probably got to a point where there is nothing to be done about sea level rise apart from being prepared to move" (I-3: 4). However, whereas other interviewees would associate this need to abandon coastal settlements with a loss of local distinctiveness, I-3 sees this solution as the logical response to coastal erosion and would not hesitate to do so. Although appreciating the attractiveness of the landscape, also I-4 criticizes the strong prevalent focus on preservation:

> "Some people get obsessed with preventing change. We can't stand still, the landscape will inevitably change. [...] I think the inherent falling back upon the Cornishness is an impediment to doing something about climate change because we won't accept the big engineering solutions that we'd need for a change." (I-4: 26)

This sober and utilitarian perspective on Cornwall's landscapes results in the demand for addressing climate change through large-scale mitigation infrastructure such as wind and solar farms on the one hand and through effective flood alleviation measures on the other; engineered if necessary. The notion of taking whatever measures necessary to adapt to the changing climate is supported by I-3 who argues that

> "we undoubtedly live in climatically changing times and we've got to be adaptable. That's how species survive, by being adaptable. That's how we transform. If you're not adaptable, Charles Darwin has got a very good description of what's going to happen." (I-3: 7)

When asked what role climate change plays in his work as a Councillor in Cornwall Council, I-4 confirms the lack of exactly such determined political will and leadership within the area's government bodies to actively implement a transformative response to the changing climate (I-4: 12) (see 5.4.1). In his eyes,

> "the landscape is under more threat from our inaction than from anything. [...] There will be more significant changes which will be brought upon us by Mother Nature really, because we failed to solve the problem and we brought environmental degradation." (I-4: 28)

In that sense he suggests that, although especially the policy narrative underlines the importance of (green) economic growth in Cornwall, this protective policy imperative is much rather a threat to a sustainable regional development than a driver of the same.

All in all, the climate change adaptation approach in the FL narrative is what one could call a "whatever it takes" approach, tolerating hard engineering solutions such as renewable energy infrastructure (in the case of mitigation) or larger-scale concreted flood defenses. The visual beauty of the landscape, even if appreciated personally by the interviewees, is viewed as secondary to the solutions necessary for maintaining functioning agriculture and communities in Cornwall.

5.5 Preliminary Conclusions

The five co-existing narratives presented above show that varying understandings of the Cornish landscapes result in divergent demands for climate adaptation approaches. It becomes clear that the concept of landscape, even if superficially understood as the same thing, has contrasting meanings for different actors. The farmed fields and hedges that are perceived as wildlife habitats in one narrative, for instance, are viewed as a food factory or evidence of ancient human settlements in others. This shows that the classification of landscapes as 'natural' or 'cultural' is highly subjective and depends on the perspective from which they are viewed. Thus, the narratives presented above correlate with the contrasting approaches to climate change adaptation in the region. Table 9 summarizes these different adaptation approaches and contrasts them with each other by listing the key elements of each landscape narrative, demands for adaptive action resulting from this, and scale on which such measures should be carried out.

In addition to these landscape narratives, the data from the first empirical phase reveals interesting insights into how the interviewees experience their individual relationships with the local landscapes. Shaped by a number of different factors such as childhood memories or leisure time activities, each of the participants show to have their own, intimate relationship with the Cornish landscapes.

In the next chapter, I highlight how the different narratives and personal accounts of the landscape – with contextualized nuances – play out in a particular location which is affected by the changing climate through coastal erosion, and where a variety of landscape management organisations are in debate over how to respond to these changes: the example of Godrevy headland. Presenting how the four narratives (excluding the policy narrative) are reflected in the viewpoints of the actors involved in this process at the local level, I show what locally-embedded discussions around landscape management in times of a changing climate arise, and how a constructivist landscape perspective can help to understand the concrete physical-material implications of the different viewpoints in this debate about climate change adaptation.

Table 9:   Comparison of Narratives' Implications for Adaptation

Landscape Construction	Focus of Adaptation	Central Actors	Scale of Adaptation Measures
The Official Policy Narrative			
Landscape as natural beauty *and* cultural heritage; central for regional identity; mining era romanticised	*not specified*	Cornwall Council	*not specified*
The Human-Environment Interaction Narrative			
Landscape natural *and* human-made; cultural heritage; distinctiveness of local practice; regional identity	Household level and communities; sustainable human in landscape; enhancing landscape knowledge	EA, National Trust, AONB, CCFF, Cornwall Council (partially)	Small-scale, bottom-up approach; individual properties and infrastructure; community-based actions
The Natural Systems Narrative			
Landscape as wildlife and habitats; protection from human harm; rejection of built structures; local species	Sustain. habitats; flood management; coop. with farmers to reduce runoff; working *with* nature	Natural England, Wildlife Trust, CoaST, Cornwall Council, EA	Individual site level to landscape scale; catchment approach to flood management
The Visual Beauty Narrative			
Landscape as beautiful natural environment; unspoilt; local distinctiveness of building materials, coastline	Protection from built structures; maintaining wild 'feel'; aesthetics of adaptation measures	AONB, Cornwall Councillor, National Trust, CoaST	Landscape shall remain scenic and attractive from afar and at individual sites
The Functional Landscape Narrative			
Production and extraction of energy; no romantic notions; engineered structures not disturbing	Adaptation of farming practice and water mgmt.; flood alleviation; renewables; efficiency of measures	Parish Councillor/Farmer, Cornwall Councillor	Individual farm level to larger areas, e.g. in case of flood management or renewables infrastructure

Source: interview data

6 Coastal Erosion at Godrevy – Adaptation in the Context of Local Debates

In chapter 5, I presented four co-existing narratives about Cornwall's landscapes and the ways in which they influence different actors' approaches to climate adaptation in broader terms. The Godrevy case study presented in this chapter draws a more concrete picture of how physical adaptation activities are embedded in local contexts and perceived differently by the actors involved. In line with a grounded theory-led approach to data analysis, the choice of this smaller-scale case study analysis resulted from the first set of interviews in Cornwall; interviews in which the participating local actors clearly identified Godrevy as the first example of conflict between the regional landscape management organisations when it comes to adapting to the impacts of climate change (cf. section 5.3.1).

In the following sections, I firstly introduce the Godrevy case and the discussions about adaptation to coastal erosion in the area and give an overview of the relevant local actors (6.1). At Godrevy, the interview data and document analysis show clearly that the translation of adaptation strategies into local actions is shaped by both normative considerations in the form of co-existing landscape narratives and different perceptions of climate change and coastal erosion (6.2) and practical considerations such as local ownership structures and the question of who should have access to the landscape (6.3). I will expand on these normative and practical considerations throughout the following section to then synthesize them and illustrate, using maps, the concrete viewpoints of the different actors on how the Godrevy landscape should be reshaped in response to the eroding coastline (6.4).

The Context of the Godrevy Case Study
As outlined in chapter 1.3.3, Godrevy is a coastal headland on the Atlantic coast of Cornwall and an important visitor attraction in the region due to its iconic scenery and diverse offers of outdoor activities (National Trust 2016b, Cornwall AONB Partnership 2016a, Visit Cornwall 2017a). It is described by the local actors as a visually attractive and iconic headland with a long coastline, vast beaches, steep cliffs topped by green heathland, and with Godrevy lighthouse as a focal point in the landscape (see Figure 16). Referred to as particularly attractive are also the rugged and dramatic rocks, the views along the north coast from the east side of the headland, the beaches, as well as the sand dunes stretching across the bay through to Hayle (I-G1, I-G3, I-G5, I-G7). I-G1.1 (TP) highlights the diversity of the Godrevy landscape and its attractiveness:

© Springer Fachmedien Wiesbaden GmbH, part of Springer Nature 2019
V. Köpsel, *New Spaces for Climate Change*, RaumFragen: Stadt –
Region – Landschaft, https://doi.org/10.1007/978-3-658-23313-6_6

"I think it is a really interesting place. Kids love it because of the rock pools and also the surf. [...] I mean every level you've got. If you're into history... It ticks all the boxes, doesn't it? Walking. Flowers. Birds. Seals. Archaeology. Just playing on the beach. Scenery, light, painting." (I-G1, 1:17:30)

Godrevy is described by all interviewees as a place with significance exceeding the local level. With over 60.000 people living in the neighbouring towns of St Ives, Hayle, Redruth, Cambourne and Poole, it is an important outdoor destination for locals. Moreover, Godrevy Lighthouse is allegedly written about in Virginia Woolf's famous novel "To the Lighthouse" and is thus known across and beyond the UK (I-G1, I-G3). Additionally, Godrevy is one of the National Trust's most popular sites in Cornwall and a destination for hundreds of thousands of visitors each year from Cornwall, the rest of the UK, and abroad (National Trust 2016b, Visit Cornwall 2017a) (cf. section 1.3.3). The interviewees underline its significance for the local economy and tourism industry, for wildlife protection, people's health and wellbeing, recreational use as well as surfing (e.g. I-G9, 1:05:03).

Figure 16: The Lighthouse viewed from Godrevy Beach
Source: own photo, 2016

When it comes to changes in the Godrevy landscape, all interviewees direct attention to the increasing numbers of visitors to the headland and, in close connection to this, the rising number of cars driving across the site every day. Likewise, an expansion of the caravan parks surrounding the headland is noticed, often connected with concern about the fast increase in new built development (I-G1, I-G2, I-G11). What problems these rising numbers of people bring and how this should be handled in terms of landscape management, however, is viewed quite contrastingly from the perspectives of the different landscape narratives (cf. section 6.2).

Landscape Management and Coastal Erosion

Godrevy headland is owned and primarily managed by the National Trust (hereafter: the Trust), but besides the Trust a number of other landscape management organisations with different foci have a stake in Godrevy. Although not including any settlements, Godrevy headland is equipped with infrastructure in the form of a café, an access road, a coastal path, and two car parks; a smaller one next to the café and a large one on top of the headland (see Figure 18, next page). In recent years, accelerating coastal erosion of the soft rock cliffs around Godrevy has threatened to undermine this infrastructure (see Figure 18, white arrows). Cliff falls have cut off a number of beach access points, and the only access road to the headland is currently less than 0,5m away from the edge (see Figure 17). With annual erosion rates of an 0,5m this road, which is running parallel to the South West Coast Path on the site, is likely to be undermined within the next 4 years (I-G3, I-G5, Earlie 2015, National Trust 2017e). In response to these developments the Trust, as the main land owner and manager, is planning to relocate the big car park, coast path and road to locations further inland (I-G3, I-G5, ibid).

Figure 17: Godrevy access road and coast path
Source: own photo, 2016

Framing coastal erosion in the area as clearly climate change-related, the Trust's adaptive activities are shaped by their central policy document "Shifting Shores", which argues for "living with a changing coastline" by working with natural processes and relocating structures away from the cliff edge wherever possible (National Trust 2008: 1). The South West England version of Shifting Shores identifies Godrevy as a primary example of a site where the changing coastline, accelerated by changing climate, impacts on the safety of public access and local facilities. As several ownership structures and landscape designations overlap at this site, the Shifting Shores document argues that "any future management options will be heavily dependent on the Trust working with others to develop a shared approach to future management and access" (ibid: 8).

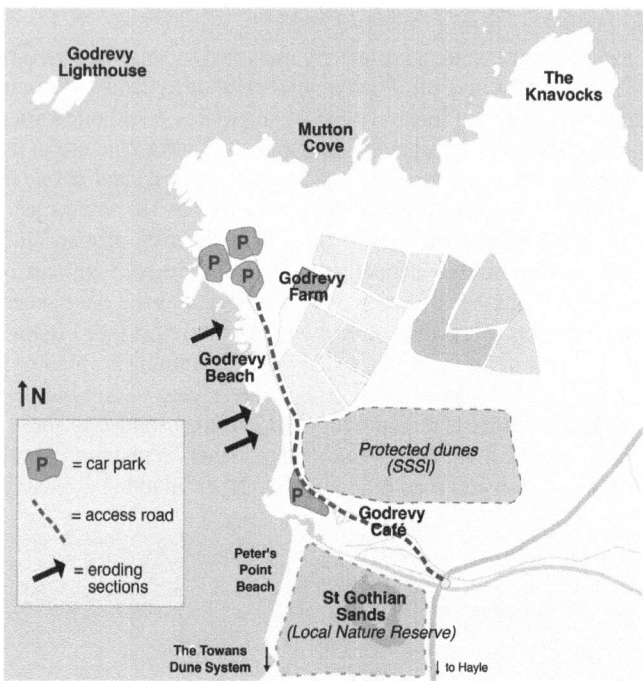

Figure 18: Current infrastructure situation at Godrevy headland
Source: © OpenStreetMap Contributors (www.openstreetmap.org), modified; interview data

To relocate the infrastructure at Godrevy, even if owning the land, the National Trust is required to submit a formal planning application to Cornwall Council (I-G3, I-G5). According to the UK's National Planning Policy Framework, "[e]arly engagement has significant potential to improve the efficiency and effectiveness of [a] planning application", and "[t]he more issues that can be resolved at pre-application stage, the greater the benefits" (Department for Communities and Local Government 2012: 45). Currently in this informal pre-application stage, the Trust is thus working towards finding a compromise between all involved actors about the new location of the car park and access road prior to entering the formal planning process. These informal discussions have been taking place for the past decade under the lead of the National Trust; despite the length of this process, however, agreement on the relocation activities could not yet be reached (see e.g. I-G3, I-G5). The Godrevy case hence constitutes a multi-actor setting where differing perceptions of the local landscape, climate change and coastal erosion shape the decision-making process about physical-material adaptation to a changing coastline. Whereas at first glance Godrevy appears to be a relatively undeveloped piece of land with many potential locations for a new road and car park, the complexity of the discussions about the relocation activities becomes visible only in dialogue with the local actors. To understand their different and often quite contrasting viewpoints in these local discussions, it is vital to comprehend the normative and practical considerations that underlie their arguments and approaches to landscape management at Godrevy. I therefore present in the first half of this chapter how the different landscape narratives outlined in chapter 5 unfold in the Godrevy case, and what implications these constructions have for landscape and adaptation management in the area. Before going into detail about these different perspectives on the Godrevy landscape and their implications for climate adaptation, the next subchapter describes the constellation of actors involved in landscape management at Godrevy.

6.1 The Actor Constellation at Godrevy

Due to the variety of overlapping landscape designations at Godrevy, a range of actors are involved in land management on the headland; each with different goals, statutory powers, and roles in the discussions around relocation (see Figure 19). Although the power relations between these actors and their practices of negotiating the relocation activities are not the primary focus of this study, it is important to understand their rights and responsibilities with regard to decision-making, and the temporal and spatial scales of their management approaches (see also Figure 20). A more in-depth analysis of their distinct viewpoints on Godrevy and the relocation activities then follows in sections 6.2-6.4.

The National Trust (NT)
The National Trust has been owning Godrevy headland since the 1950s (I-G5). The organisation is leading the decision-making process around the relocation activities and is responsible for implementing and financing any measures. Its responses to coastal erosion are importantly influenced by the Shifting Shores strategy (National Trust 2008: 2). This document identifies the key principles of adapting to a changing coastline to be long-term planning, supporting and working with natural processes, considering the wider context of the managed sites, finding solutions in partnership with other involved actors, and enabling participation by the public.

The slogan of the Trust is "Forever, for everyone", implying a long-term perspective on planning and strategy-making (National Trust 2017a) (cf. section 1.3.1). Embedding their landscape management activities into national policy

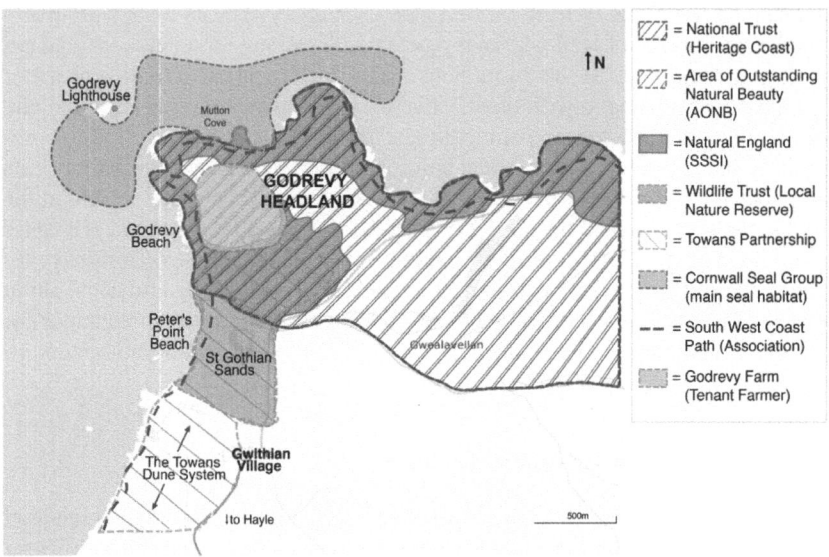

Figure 19: Overview Landscape Management Organisations Godrevy
Source: © OpenStreetMap Contributors (www.openstreetmap.org), modified; interview data

frameworks such as "Shifting Shores" and other organizational guidelines, the Trust draws on a broad basis of national-level policy and strategy documents in their decision-making process (National Trust 2014, 2015, 2017a). Striving for local agreement about the new location of the car park and road *before* the formal planning process, however, the organisation refrains from taking a decision about adaptation in their own hands and acknowledges that "there are many people involved, and there are lots of different interests" among the actors on-site (I-G3, 19:02) (see above). As the Trust is aware of other actors' formal veto rights in the planning process, its local staff seek to maintain positive relationships with all neighbouring organisations (I-G3, 37:02). Nonetheless, the particular view-point of the Trust on landscape management stands in contrast with those of a number of other actors at Godrevy (cf. sections 6.2.-6.4.).

AONB Partnership (AONB)
Godrevy headland is part of the "Godrevy to Portreath AONB" (Cornwall AONB Partnership 2016a). Usually working together with the "Trust [as] one of [their] partners", the manager of the AONB Partnership describes the Godrevy case as "a good example of the potential conflicts that we're looking at" in the context of climate adaptation in Cornwall (I-18: 13). The strategic goals of the AONB

are formalised in the form of management plans every five years that cover topics such as 'landscape & seascape', 'climate change & energy', and 'community & economy' (Cornwall AONB Partnership 2011b). While the National Trust is leading the discussion about relocation at Godrevy and will file the formal planning application, the AONB Partnership has a statutory veto right and can legally object to this application during the official consultations (I-G3, I-G5, I-G11) (Department for Communities and Local Government 2012, Cornwall AONB Partnership 2017).

Towans Partnership (TP)

In charge of managing the Towans[18] dune system, which borders Godrevy headland in the south west, the Towans Partnership is a well-established local association of actors with interests in the Towans. Engaged in the Partnership are local businesses, Cornwall and Parish Councillors, the Wildlife Trust, local police, and Natural England (Towans Partnership 2016a). The Partnership was "established [...] to focus on facilitating communication between different agencies and groups" (ibid). Not having any statutory veto rights in the planning process, they describe themselves as a "useful umbrella group, a forum" linking the different actors in the Towans area (I-G1, 11:50). The Partnership is a well-known local body with a strong standing in the area and is thus actively involved in the discussions with the Trust (I-G1, I-G3, I-G5). The partnership's management focus is solely local and primarily reactive regarding both visitor management and responses to coastal erosion (I-G1, Towans Partnership 2014).

Natural England (NE)

As "the government's adviser for the natural environment", Natural England is responsible for the management of the "Godrevy Head to St Agnes Site of Special Scientific Interest (SSSI)", part of which is the protected dune grassland south of Godrevy farm (Natural England 2017b). Having a "statutory responsibility for the landscape", Natural England can obstruct the Trust's plans in the formal planning process should they interfere with their goals of protecting and enhancing the state of the SSSI (I-G4, 10:20, Department for Communities and Local Government 2014). Natural England operates at a similar spatial and temporal scale as the National Trust: their work is embedded in a national organisational and policy framework, drawing on a wide context of management plans and strategies (I-G5, see e.g. National Trust 2015, 2017c).

---

18   from Cornish: *tewyn,* meaning 'sand dune' (CORNISH LANGUAGE PARTNERSHIP 2017)

Cornwall Wildlife Trust (CWT)

The CWT describe themselves as "the leading local charity working to protect Cornwall's wildlife and wild places" to conserve habitats and protect wildlife (Cornwall Wildlife Trust 2017a). At Godrevy, they are responsible for the Local Natural Reserve in the Upton Towans, which neighbour Godrevy headland (I-G2, 12:20). Responsible for this Nature Reserve, the Wildlife Trust is in a position to object to the Trust's plans if they interfere with habitat protection (I-G2).

Cornwall Seal Group (CSG)

The Cornwall Seal Group is a "conservation charity supporting a large network of active citizen scientists" and researching the colony of Grey Atlantic Seals at Godrevy (Cornwall Seal Group 2017). Godrevy is one of two major haul out sites in the entire South West of England for these seals and thus is the main focus of the CSG's activities. Being a voluntary organisation, Cornwall Seal Group holds no legal powers whatsoever. The decision about the relocation of the car park, however, would importantly affect their efforts to protect the sensitive seals, giving the group a particular interest in the outcomes of this discussion (I-G9, 41:20).

South West Coast Path Association (SWCPA)

Although also not actively involved in the discussions around relocation at Godrevy, the Coast Path Association is responsible for managing the coastal paths along the coast of South West England (South west Coast Path Association 2017). At Godrevy, this path is significantly threatened by the ongoing coastal erosion (I-G10).

Visit Cornwall (VC)

Offering Godrevy as "the ultimate outdoor experience", Visit Cornwall, the Cornish tourism board, is one of the most important bodies in Cornwall's tourism sector and features the headland in their advertisements of the area (Visit Cornwall 2017a).

Surf School (SS)

The "Shore Surf School" is a privately owned business near Godrevy Beach. Being originally from Cornwall and having operated the business since 2000, the owner of the school actively uses Godrevy headland as the basis for his livelihood and is dependent on visitors coming to the site. Although he has little negotiation power in the discussions about relocation, he holds intimate experiential knowledge about the Godrevy area and its beaches (I-G12, Shore Surf School 2017).

Tenant Farmer (TF)
The local farmer holds a tenancy agreement with the National Trust and has cultivated the arable land at Godrevy for the last 25 years, also living in the farmhouse on the headland until recently (I-G5, I-G6). Due to his tenancy he is, on the one hand, part of the braid of ownership structures in the area that complicate the relocation activities (cf. section 6.3). On the other hand, he is directly affected by the Trust's decisions around land management (I-G6).

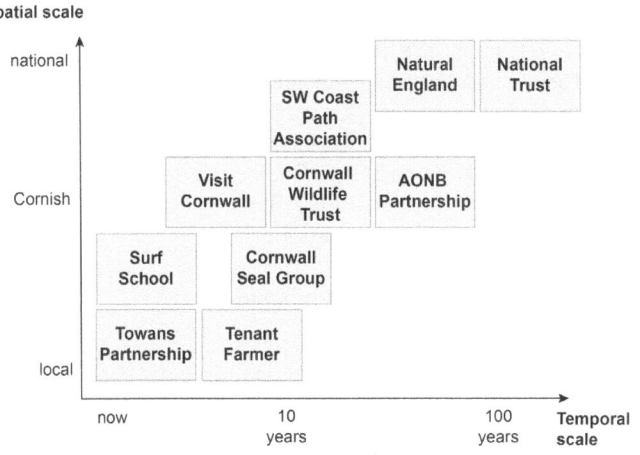

Figure 20: Scales of landscape management at Godrevy
Source: own figure

After having introduced the different actors involved in managing the Godrevy landscape, in the following subchapter I present the normative considerations that influence their decision-making around the relocation activities. These normative considerations are on the one hand the different narratives about the Godrevy landscape and its management, and on the other the differing perceptions of climate change and coastal erosion.

6.2 Normative Considerations: Landscape Narratives at Godrevy

Similar to the narratives about Cornwall's landscapes in general (cf. chapter 5), different co-existing constructions of the landscape emerge from the interviews with the actors at Godrevy. These constructions are embedded in a specific local context and have concrete implications for landscape management in the area. Although the interviewees have certain shared perceptions of the headland (cf.

introduction to this chapter), when entering into dialogue with them, a very nuanced picture of their constructions of the local landscape comes to light. The narratives of the landscape as human-environment interaction, natural systems and a functional space for production are reflected in the interviewees' perceptions of Godrevy headland (cf. chapter 5). However, whereas most actors agree on the high symbolic significance of the site in its wider regional context, within those narratives the interviewees have quite different understandings of the landscape itself. This leads to the co-existence of diverse management approaches, viewpoints on climate change, and responses to the accelerating coastal erosion. Moreover, comprehending the lines of argument throughout these narratives is an important basis for understanding the debate about the Godrevy landscape as a common good and the question of access to it in times of an eroding coastline (cf. section 6.3). For an overview, Appendix 8 highlights which narratives each of the interviewees at Godrevy follow. Table 10 at the end if this section summarizes the different constructions of the Godrevy landscape and contrasts their key aspects and goals of landscape management.

6.2.1   Godrevy as Human-Environment Interaction

From the HE perspective, the Godrevy landscape is constructed as a melding of natural and human-made features with a high degree of visual attractiveness and as a popular destination for visitors enjoying the outdoors. This perspective is most strongly represented by the National Trust as the main land owner at Godrevy, but also by the AONB Partnership, Visit Cornwall and others. Whereas all these actors agree that the landscape consists of *both* natural and anthropogenic features, their foci of landscape management nonetheless differ.

Godrevy is presented by the Trust as an iconic and attractive coastal headland of national importance which is a magnet for visitors who enjoy the outdoors and the local beaches (I-G3, I-G5,). The main leisure activities undertaken onsite are walking with and without dogs, surfing, water sports, enjoyment of the beaches, and watching local wildlife (I-G3, I-G5) (see Figure 21). Godrevy's significance for Cornwall and the UK is thereby understood in terms of what it means to the people who visit the site: long-standing attachments to it as a holiday destination and family histories written on the local beaches (I-G3, I-G7). The use of the beaches by the surf community creates an element of "real local ownership" of the landscape (I-G3, 06:10). In this sense, the Trust constructs Godrevy very much as a commons, a place the main purpose of which is to be visited and enjoyed by people (I-G3, I-G5) (cf. section 6.3.1). Whereas the protection of the local wildlife and healthy habitats are among the foci of the National Trust, their main goal is to manage the landscape *for* people (I-G3, I-G5).

Figure 21: Portrait of Godrevy by the National Trust
Source: ©National Trust Images/Ben Selway

Major changes perceived by the staff of the National Trust at Godrevy are the increase in visitor numbers and the proceeding coastal erosion (I-G3, I-G5). The Trust's "dilemma really" when it comes to managing access to the landscape is that Godrevy "has become more and more popular over the years. And a lot of the natural aspects of the place are suffering" (I-G5, 02:10). The reason for these changes are viewed to be a combination of the large numbers of visitors to the site, its location and soft rock geology, and an increase in wave action and heavy storms (I-G5, 12:00). For the Trust, landscape management "fundamentally […] is around managing access and the numbers of people who want to come here whilst trying to maintain the natural beauty of the site." (I-G5, 02:10). Their decision-making is importantly driven by the wishes of those who visit the site: their members, the local communities and tourists, but also the staff of other organisations involved in managing the headland. However, although providing access for people is the main goal of the Trust, the high volume of cars driving across the headland constitutes a particular management challenge as car-induced erosion is especially problematic during wet weather (I-G3, 1:06:00). It is thus seen as important to improve the visitor infrastructure whilst not obstructing the natural looks of the landscape and its ecological sustainability (I-G3, I-G5).

Also the AONB and Visit Cornwall construct Godrevy in terms of the HE narrative; however, their management focus slightly differs from that of the Trust (I-G7, I-G11). From the perspective of the AONB Partnership, the landscape is a combination of

> "land cover, topography, drainage, soils, woodland cover, cliff profile, geology […]. All of those things come together to make up natural beauty and a unique combination in a place. And that's perceived by people, and that's got an identity element to it and a cultural one." (I-G5, 47:30)

While constructing Godrevy similarly to the Trust, from the AONB's perspective a particular concern is the 'greying out' of landscape character and natural beauty due to bad management choices and the rising number of visitors (I-G7, I-G11). Especially "agriculture pinning in the costal habitat […] and Cornish Hedge removal […] erodes the character of the area" (I-G11, 20:05). When referring to the natural elements of the landscape, there is no focus on wildlife and its protection, but primarily on visual aspects (I-G7, I-G11). The current headland car park and other built structures in the area are therefore criticized for being "very, very visible" from many locations in the landscape (I-G11, 23:45). Although agreeing with the Trust that the landscape is a common good, ideal management of Godrevy hence means to "conserve and enhance natural beauty" and prioritise this goal over others goals, if necessary (I-G11, 43:35; I-G7).

The owner of the local surf school also constructs the landscape from both its natural ans human-made elements. In terms of the history of Godrevy, he points out that even the dune system of the Towans, which presently appears to be natural, was largely shaped by past human activities such as sand mining and dynamite production (I-G12, 02:20). He thus is very aware of the human impacts that have shaped today's landscape in the Godrevy area.

However, for him the natural look and "untouched" atmosphere of Godrevy play a central role in what makes the site so popular (I-G12, 02:50) (see Figure 22). Furthermore, his first experience with Godrevy was a fundamentally visual one: "A few years before we did the surf school I […] just looked over the hill and thought 'Oh my God, look at this! That's amazing!'" (I-G12, 06:15). Being originally from Cornwall and having a close connection to Godrevy through personal experiences, he has a strong emotional attachment to the landscape and appreciates its natural beauty. In terms of changes to the landscape, he refers to the coastal erosion around the headland and, experiencing the landscape on a daily basis when conducting surf lessons and spending time on the local beaches, displays detailed knowledge about the changes to beach layouts and recent cliff falls (I-G12, 15:35). He notices these increasing pressures on the headland based on the condition of the car parks and the café on the headland, again highlighting the importance of the site for tourism (I-G12, 10:25). Agreeing with the Trust on the significance of access to the landscape, for him the main benefit of visitors to Godrevy is maintaining a successful business (I-G12, 23:24).

For I-G10 (SWCPA), too, it is his daily visual experiences that shape his perception of the Godrevy landscape. Seeing the headland and the lighthouse from his home town of St Ives across the bay every day, he finds that he has "a very, very close relationship personally" with this landscape which is "part of [his] life really." (I-G10, 08:10). In his capacity as a representative of the Coast Path Association, it is the continuity of the coast path he is concerned about when it comes to landscape management. Both the erosion processes on the site and the high numbers of visitors each summer present significant pressures to the path (I-G10, 07:15). Focused on possibilities to relocate the coast path further inland, he is thus very aware of the different ownership structures at Godrevy and concerned with the more practical questions around access to the coast (I-G10, 25:30).

Figure 22: View from Godrevy headland over Peter's Point Beach
Source: own poto, 2016

Perceptions of Climate Change and Coastal Erosion
Similar to the broader narratives about the Cornish landscapes (cf. chapter 5), the different constructions of the Godrevy landscape go hand in hand with varying understandings of both climate change and coastal erosion. While all interviewees agree that erosion is a "very life issue […] at Godrevy" (I-G3, 15:55), perceptions of climate change per se and how this erosion should be addressed diverge between the actors at Godrevy; having clear implications for the discussion around the relocation activities. The interviewees perceiving Godrevy in terms of the HE narrative agree that climate change is happening in Cornwall and that human activity is accelerating it (I-G3, I-G5, I-G10, I-G11, I-G12). From this perspective, climate change impacts on Godrevy are perceived to be more frequent and stronger storms that increase wave action, more heavy rainfall events destabilizing the cliffs, and proceeding coastal erosion. A number of indirect effects result from these physical impacts of climate change on Godrevy, such as the loss of beach access through the erosion of stairs (see Figure 23), damage to of loss of sections of the coast path, changes to beach layouts, as well as the deterioration of paths across the headland (I-G3, I-G5, I-G10, I-G11, I-G12).

Figure 23: Closed-off beach access point to Godrevy Beach
Source: own photo, 2016

When it comes to responding to the climate change impacts on Godrevy, there is a shared sense of urgency among the interviewees following the HE narrative. For staff of the Trust, who speak about tackling erosion "for future generations" (I-G3, 03:30), now is the time to move from reactive decisions about climate change impacts to thinking ahead (I-G3, I-G5). Regarding coastal erosion all interviewees agree that "coastal change due to climate change is a huge issue" in the area (I-G11, 16:20) and that it needs to be addressed through adaptation measures. For I-G5 (NT), the erosion of the cliffs around the headland is "related to what we perceive as different weather patterns, more storm events" (I-G5, 13:45). Especially by the Trust it is seen to be "probably one of our most visible issues at the moment" (I-G3, 16:30). Although the erosion at Godrevy is understood as being climate change-related from the HE perspective, the interviewees also connect it to the human activities on the headland. Landscape change at Godrevy is framed as "a mix between the coastal erosion and also the lot of people coming to the site" (I-G3, 14:30). To respond to this erosion, 'soft' solutions that do not interfere with the visual attractiveness of the landscape are preferred over hard-engineered structures. I-G11 (AONB) suggests to simply "allow [the cliffs] to fall into the sea and you allow it to do so safely" (I-G11, 34:30). As becomes apparent from their policy document 'Shifting Shores', adaptation to a changing coastline is also a huge topic for the National Trust. "By recreating a naturally functioning shoreline", the document states, "we free ourselves from the 'sea defence cycle' of construct, fail and reconstruct" (National Trust 2014: 2). In practice, however, working with natural processes and relocating structures from the cliff edge is not always that easy (see also 6.3.2):

"Looking at the coastal strip around Cornwall, it's not always possible to acquire land […]. We're working with land owners and tenant farmers to try and look at the conditions of land but… There's a lot around that strategy of just trying to work with nature." (I-G3, 22:40)

6.2.2   Godrevy as Natural Systems

Whereas the HE perspective centres on people and their attachments to the Godrevy landscape, in the natural systems narrative people only play a secondary role. Here, the headland is conceptualized as a mosaic of sensitive wildlife habitats and ecosystems. This perspective on Godrevy reflects the strong notion of protection and stewardship also found in the broader NS narrative about the Cornish landscapes (see 5.3.2.). Natural England, the Cornwall Wildlife Trust, the Towans Partnership and the Cornwall Seal Group follow this narrative; however, once again with different foci and implications for landscape management.

Responsible for the Godrevy SSSI and primarily concerned with wildlife protection, the staff of Natural England perceive the site as a sensitive coastal habitat and dune ecosystem (I-G4). The elements referred to are primarily from the natural environment, and the wording circles around the local flora and fauna, e.g. the "coastal heath- and maritime grassland" and a variety of species (I-G4, 19.20) (see Figure 24). Central characteristics of the local landscape are described to be the "continuity of the system" and the "great [habitat] connectivity" (I-G4, 19:20). This landscape perception is accompanied by a great eye for detail and knowledge of local biodiversity. I-G4.2 (NE), for example, alerts her colleague to animals that she notices in her immediate surroundings during the interview: "Did you see that bird? Tiny little bird and making such loud noise! What kind was that?" (I-G2, 23:05).

From Natural England's perspective, landscape management at Godrevy is described in terms of controlled grazing, fencing, removing shrub, controlling invasive species, as well as certain practices of farming (I-G4). Importantly, Godrevy headland is perceived to be in need of protection from overuse by human activity:

"The pressures are quite extreme. People need space as well as nature, and we need to try and get the balance right so that nature is not pushed out by people." (I-G4, 20:50)

When it comes to looking after this landscape, it is important for Natural England to support natural processes. However, whereas a distinction is made here between the natural and the human system and humans are not seen as part of the actual landscape, there is recognition especially from the side of Natural England that Godrevy is a significant site not only for wildlife, but also for people to spend their leisure time and holidays (I-G4).

Figure 24: Hudsonian Whimbrels on top of Godrevy Head
Source: own photo, 2016

Godrevy is portrayed by Cornwall Wildlife Trust as a 'wild' landscape (also termed 'semi-natural habitat') that is more untouched by human activity than most other places in Cornwall. The wording here, too, centres on terms from ecology, biodiversity and hydrology such as "rare species of butterflies" and "the second largest dune system in the country" (I-G2, 11:10; 25:45). Godrevy is viewed as a nationally important coastal and dune heathland due to the absence of settlements on the headland and because of its protected sites (I-G2, 11:15). The popularity of Godrevy is perceived to be rooted in the fact that "people like to come out and enjoy fairly natural places" (I-G2, 24:30). Regarding the management of the landscape, the Wildlife Trust is mainly concerned with two aspects: the impacts of walkers and dogs on the sensitive coastal habitats in the area, and the encroachment of invasive plant species. In this context, I-G2 (CWT) expresses concern with the growing infrastructure and settlements in the hinterland of Godrevy that hinder the dunes from moving further inland, as they would naturally do (I-G2, 43:10). He would, however, "not be looking to stop people coming with their dogs, but […] would like to do something about the amount of dog fouling" (I-G2, 38:10). The protection of the dune ecology thus stands in the centre of landscape management (I-G2).

I-G9 (CSG), whose research focuses on the colony of seals in one of the local coves (see Figure 25), points out that Godrevy is "one of four really important sites in the Southwest" for the animals to spend time ashore – and is thus very important for the protection of the species (I-G9, 07:35). This fact makes Godrevy an especially sensitive wildlife site and requires special attention regarding both landscape as well as visitor management (I-G9, CSG). A particular concern for her is therefore the currently short distance from the headland car park to Mutton Cove, where the seals haul out to rest. Since the car park lies a very lively area, she argues, walking the path up to the cove does not allow people to quiet down to not disturb the seals. On the other hand, however, she acknowledges the need for people to visit Godrevy headland to educate them about the seals and make them understand the need for wildlife protection.

Figure 25: Grey Atlantic Seals at Mutton Cove
Source: own photo, 2016

From the perspective of the Towans Partnership, Godrevy is part of the second largest dune system in Cornwall which is "nationally important for its wildlife" and its "high value for habitat diversity" (Towans Partnership 2016b) (see Figure 26). I-G1.1 (TP) describes Godrevy headland as "one of the most iconic views of Cornwall" due to its coastal beauty and argues that, if one would pick ten views representative of the region, Godrevy would certainly be among them (I-G1, 1:16:40). The focus, however, clearly lies on the protection of the dune system: as stated in their Management Plan, the priorities at Godrevy are the "protection, safeguarding and enhancing" of the local landscape and habitats (Towans Partnership 2014: 8). The interviewees from the Towans Partnership hold detailed knowledge about local species, the different physical elements of the landscape and their functions. I-G1.1, for example, refers to the soft cliffs above Godrevy a "peri-glacial head" (I-G1, 45:37) and is familiar with local bird species such as Sand Martins which, as he points out, "nest here" (I-G1, 45:20). Taking visual cues while walking across the headland, he names species, explains developments in the natural environment and displays detailed knowledge about them:

> "Celandines! I was reading about them recently in 'Flora Botanica' and the name Celandine comes from a word in Greek– caledron or chaledron meaning swallow." (I-G1, 1:13:50)

Both interviewees from the Partnership express attachment mainly with the natural features of the Godrevy landscape. At the same time, concern is expressed with the behaviour of visitors and locals especially when walking their dogs, which is viewed as harmful to local ecosystems and wildlife (I-G1, 12:42). Visitors are perceived as intrusive and potentially harmful for the sensitive coastal ecosystems of Godrevy, especially when deviating from the footpaths and disturbing local wildlife such as the seals at Mutton Cove (I-G1, I-G4, I-G9). In contrast to the HE narrative, people and cars are not seen as part of the landscape here, but as an external negative impact *on* the landscape; a perspective that reflects the separation of the natural and the human system in the NS narrative (I-G1, 28:35) (see also 5.3.2). Controlling such harmful impacts on the landscape is referred to by the Towans Partnership as "people management" and is – besides the management of coastal erosion – highlighted to be their key tasks at Godrevy (Towans Partnership 2014: 15).

Figure 26: Godrevy as important wildlife habitat
Source: Towans Partnership 2014: 8

Perceptions of Climate Change and Coastal Erosion
From the NS perspective, climate change at Godrevy is perceived as a process which "has always happened anyway" and as part of a larger natural circle of events (I-G1, 40:01). It thus clearly mirrors the viewpoint of the broader natural systems narrative (comp. 5.3.2). For I-G9 (CSG), moreover, "it doesn't really matter why it's happening. It's a fact that things are changing" (I-G9, 04:40). The interviewees from the Towans Partnership admit that they are "not really talking about climate change on the ground" as there is still a "degree of uncertainty around what is happening and what are the impacts" (I-G1, 48:00; similarly I-G2, 18:30). Increased storminess and rainfall as well as changes in sea and air temperatures are described as the direct climate change impacts on Godrevy (I-G1, I-G2, I-G4, I-G9). The encroachment of invasive species, changes to local wildlife habitats, effects on the natural movement of the dune system, and coastal erosion (I-G1; I-G2, TP) are all perceived as resulting from these impacts. Having researched the local seal colony for the past two decades, I-G9 (CSG) noticed significant changes in the mammals' behaviour and argues that "the obvious reason why seals would be confused is because of climate change" (I-G9, 21:50).

Moreover, for I-G2 (CWT), the changes brought about by a changing climate will affect the entire sensitive dune ecosystem of the Towans:

> "Sand dunes [...] are a dynamic system, they're changeable. [...] And with climate change, if we're going to have wetter winters when we have these stormy high winds, than there might not be as much mobile sand available." (I-G, 24:00)

All interviewees with a natural systems perspective on Godrevy confirm that coastal erosion is happening at Godrevy headland at an increasing rate. They clearly connect "climate change pressure [...] with potential coastal erosion acceleration" (I-G9, 03:26). An exception is the Towans Partnership: I-G1.1 doubts that "many would link what is happening here to climate change, the erosion" (I-G1, 41:20). While coastal erosion is mentioned as an ongoing process at Godrevy on the Partnership's new visitor information boards in the dunes, climate change is not mentioned as a potential cause (Towans Partnership 2016b). It is generally agreed upon in the natural systems perspective, however, that costal erosion is a natural and positive process that is not necessarily a reason for concern. I-G1.1 (TP) argues that "we would say this is creating sediment down there for the beach! It's all natural" (I-G1, 45:20). Although their organisation acknowledges coastal erosion as a "threat to infrastructure, access roads, cap parks", it is understood as a process that will "introduce further valuable habitat diversity, and should be encouraged wherever possible" (Towans Partnership 2016). When it comes to managing the landscape in the context of coastal erosion the use of natural, non-intrusive solutions is clearly preferred over engineered structures from the natural systems perspective (I-G1, I-G2, I-G4). I-G2 (CWT) argues that "not putting in any sort of hard sea defenses is very sensible" to not "ruin the scenery and change the dynamic system processes" around the headland (I-G2, 33:30). I-G9 (CSG) finds that "there isn't an awful lot else we can do about that" other than working with the natural processes (I-G9, 41:00), however, the staff of Natural England argue that:

> "We have an understanding and an appreciation that there needs to be adaptation. You can't hold the line everywhere. And so we need to work out what would the most beneficial scheme which we would be able to support." (I-G4, 04:40)

Bringing climate change into the discussions around the relocation activities at Godrevy, I-G4 (NE) explains, "immediately lengthens the time frame" of their planning activities and evokes more strategic thinking about landscape management.

6.2.3   Godrevy as a Functional Landscape

The functional landscape narrative from the previous chapter is also featured among the different perspectives at Godrevy. It is represented by I-G6 (TF), the tenant farmer cultivating the fields in the center of the headland (see Figure 27), who has a similarly utilitarian perspective on the landscape as a number of interviewees from the first field phase (see 5.3.4). Merely viewing Godrevy as a functional space for agricultural production, his landscape construction is lacking any notions of preservation, protection, or scenic beauty:

> "It is a coastal farm. Very sandy soil, because obviously near the coast […], and it's now very high PH levels. So this is our style of farming. Just concentrating on broccoli in the summer and cauliflowers in the winter." (I-G6, 02:20)

The wording in which he describes Godrevy circles around soil quality and PH levels, cropping and planting, harvesting technology and different uses of pesticides (I-G6). Whereas other interviewees – especially from the NS narrative – perceive farmland as being of low biodiversity value and disrupting habitat connectivity, I-G6 (TF) points out that

> "in the spring we see lots of those little Whitetails [birds] […] and we've got sparrows on the hedges, lots of swallows and swifts swooping down and around for the little insects. […] The hedges got lots of flowers and things on. So we're farming in a modern way, but I'd like to think it's sympathetic towards Mother Nature. Because I can't afford to upset Mother Nature." (I-G6, 35:20)

This statement not only shows how differently he perceives the fields he is cultivating, but also reveals his conceptualization of nature. Subjectivising nature, he understands her as a force he depends on with his business and which is outside the reach of his influence.

The changes I-G6 perceived at Godrevy in the past decades are related primarily to changes in the agricultural sector and new approaches to farming, especially "because we get modern technology" for planting and harvesting (I-G6, 10:40). In addition, he observed that the site has become busier and visitor numbers are "unbelievable" now, especially on sunny days (I-G6, 05:50). Landscape management from this functional perspective centers on improving processes of planting and harvesting and the question of which fields are suitable for which crops. Skeptical of a protective approach to landscape management, I-G6 criticizes the urge to put the entire headland under nature protection:

> "They've got the dunes there which are 102 acres and there's lots of different [species] up there. You've got The Knavocks up there which is 98 acres where they've got loads of species. […] You've got all the hedges on the farm. What more do you want? I mean, we've all got to eat." (I-G6, 20:10)

Figure 27: The tenanted farm in the center of Godrevy headland
Source: own photo, 2016

Perceptions of Climate Change and Coastal Erosion

Following the FL narrative, I-G6 (TF) perceives climate change at Godrevy mostly in terms of changes in the weather patterns that he noticed based on farming practice. He finds that especially "last winter was a weird winter, never seen one like that" (I-G6, 07:35). Although he thinks that winters like the past one could potentially be connected to climate change, he argues that "1976 was a drought, […] 1977 was wet. […] 1984 was dry – so we've seen lots of different things" (I-G6, 09:10). Whereas the other interviewees mostly speak about climate change from the perspectives of their organisations' policy backgrounds, I-G6 clearly draws on experiential knowledge gathered during two decades of farming Godrevy headland. He is "not denying climate change, but Mother Nature does change and does funny things that we cannot explain" (I-G6, 29:24). Again displaying an almost fatalistic view on nature, he thinks that

> "the cliffs will fall away, but what can we do? We can't really do anything. The rock is so soft. So […] it's all down to Mother Nature really" (I-G6, 40:10)

Despite his skeptical outlook on the changing climate, I-G6 confirms that coastal erosion at Godrevy is occurring and that especially the winter of 2013/14 "probably did more [damage] than the other 24 years" he witnessed at Godrevy (I-G6, 26:52). Hoping, however, that this winter was a "one off" (I-G6, 26:56), his perception of his own role in adapting to climate change is similarly pessimistic. Feeling like "it's just our leaders of our countries" who can respond to the changing climate because "they are the ones with the power" (I-G6, 45:03), he does not see much need for action on his own part.

6.2.4 Individual Relationships with the Godrevy Landscape

While the four narratives about Godrevy outlined above constitute collective, sub-societal constructions of the local landscape, the present study also gives insight into the individual, personal accounts of the landscape by the interviewees. In terms of individual landscape preferences, the interviewees agree that there are certain qualities of the Godrevy landscape which "kind of take a hold on you that is hard to describe" (I-G3, 08:31). All interviewees agree that Godrevy is a place of remarkable natural beauty where they enjoy spending time professionally and privately. Having visited the site for conducting research every week since 1999, I-G9 (CSG) has a very close personal relationship with Godrevy and describes the headland as her "heaven on Earth" (I-G9, 11:05). The aesthetics of the landscape strongly appeal to her personal preferences and thus shape her attachment to Godrevy:

> "I'm not religious, not very spiritual, but it's quite a spiritual place for me with very important personal meaning. […] It must just be some landscapy thing. Something about beauty and aesthetics." (I-G9, 12:30)

For her, Godrevy is "quite a spiritual place […] with very important personal meaning" (I-G9, 07:30). In a similar notion, I-G3 (NT) connects his particular attachment to Godrevy with different features of its landscape such as the "vast expense of ocean" all around the headland and the "really beautiful wildlife" (I-G3, 08:30). Even after having worked on the site for nearly 20 years, his colleague I-G5 (NT) also still finds amazement in the aesthetics of the landscape "still[s] get a buzz out of the place, because it has got natural beauty" (I-G5, 04:50). For I-G1.1 (TP), the appreciation for the beauty of the site also plays a central role in his professional life:

> "You're out, you're getting the fresh air, enjoying the light... You know, it's everything! I mean... It's my office! [...] What I mean by that, because I work outside a lot... I had an office job last year and that lasted three weeks, because it just killed me. I have to be outside." (I-G1, 34:20)

Closely connected to these personal preferences and the appreciation of Godrevy's attractive natural features are the interviewees' experiences of the landscape in different kinds of weather. While I-G6 (TF) finds the "micro-climate" on the headland to be mild due to a constant "breeze […] coming off the sea" (I-G6, 09:35), I-G11 (AONB) is impressed by the surprising wave action that can occur at Godrevy in bad weather, which gets "really dramatic […] during storms" when the waves "really, really pound" (I-G11, 14:35). Looking out to sea and pointing at the lighthouse, I-G12 (SS) recounts that he enjoys the landscape most in dramatic weather conditions:

> "There are certain days when the sea is so rough that you've got spray going over the lighthouse! There's an outer reef where that boat is over there [...], and the waves breaking out there, that's just phenomenal. It's a totally different landscape then." (I-G12, 32:20)

While most interviewees who relate to the weather highlight their preference for stormy and rough conditions (see also 5.3.5), I-G9 (CSG) especially likes Godrevy on days like the one she describes. Her statement connects both her personal and professional interest in Godrevy and mirrors her passion for the local wildlife:

> "I think I've potentially had my best day ever here last Monday. […] The sea was flat calm, there was no wind - that's quite unusual, no wind! - and it was sunny. […] I was surveying those seals on the bottom and on the island, and then two dolphins came through from the Channel. It was awesome!" (I-G9, 06:20)

Whereas primarily connected to Godrevy through their professional positions, many of the interviewees of the second empirical phase also have close relationships with the local landscape through their private lives. Spending relevant amounts of their leisure time around the headland, a number of interviewees know the landscape from the perspective of surfers. They thus not only manage the local beaches, but also use them as visitors and engage in one of the central elements of Cornish lifestyle in the outdoors. I-G2 (CWT), for instance, frequently "used to surf here and body board here" (I-G2, 05:55). I-G11 (AONB) relates to the landscape through the colours of the ocean water and the visual beauty of the site, displaying her own experiences with the landscape:

> "Look at the sea! […] I love the surf here. […] It's really good wind here today, nice surf. And really lovely […] waves. I like that. […] It can get really Caribbean here. It is gorgeous, actually." (I-G11, 08:40)

I-G3 (NT) made his own very personal experiences with the wildlife of Godrevy, which for him is part of the landscape, when he went surfing recently:

> "We were surfing here last year, and a seal popped up in front of us and looked our way. And it was one of those moments when I was just in absolute awe. You just feel at one with the nature around you and the landscape!" (I-G3, 08:40)

With this statement, he picks up the notion of therapeutic landscapes from the broader visual beauty narrative: the landscape as a natural environment which, with its natural elements and aesthetic characteristics, has the potential to feel at balance with oneself and leave the stresses of daily life behind (comp. 5.3.3). I-G12 (SS) even incorporates these landscape qualities into his work routine: "When it goes quieter, you know, I'd do that in the morning. I would just sit against the life guard hut and do some work on my phone" (I-G12, 39:35).

The fourth way in which the interviewees relate to the Godrevy landscape is through personal experiences from the past which tie them to the place emotionally beyond the professional realm. Speaking about his relationship with Godrevy before he was managing it for the National Trust, I-G3 recounts that to him and his wife, the site has "always been one of our special places"; a fact which strongly motivated him to apply for an open position at Godrevy when he saw it advertised (I-G3, 07:55). Having grown up in the area, I-G7 (VC) vividly remembers "hanging out" on the beaches around the headland and drinking there with friends during his time at university (I-G7). Also for I-G12 (SS) Godrevy inseparably intertwined with his private life. He and his business partner have

> "developed a real love for this area. And [...] in this area I met my missus. She learned to surf at the surf school and now loves the place. She runs a café a little bit further down the beach. [...] And I also live very locally now, as close as I can." (I-G12, 07:55)

Also I-G5 (NT) has a long-standing relationship with Godrevy which blurs the boundaries between his private and professional life. Having been the main manager of Godrevy for 15 years before moving up in the ranks of the National Trust, he used to regularly visit the site with his family. In recent years, however, he found that through his work he got

> "involved with it in many more ways than normal visitors would. I guess nowadays I rarely visit it as a visitor. [...] I brought my kids down here to play on the beach, had days out, 'Let's go to Godrevy'. I rarely do that now. [...] You don't always want to mix business and pleasure really." (I-G5, 04:00)

Whereas most of the interview partners associate positive feelings with Godrevy, the personal experience of I-G2 (CWT) of the site is tainted by the difficulties he has with it professionally. As he is lacking time and funding to properly manage the Local Wildlife Reserve at St Gothian Sands currently, he has "at the moment a bit of a negative vibe about it" and admits that hence, he was "a little bit nervous about this interview" (I-G2, 02:30). His statement – along with all the other accounts of individual relationships with Godrevy – shows clearly that, even though interviewed in their roles as organisational staff, the interviewees' constructions of the local landscape meld their personal *and* professional perceptions of the place.

Summarizing the Different Landscape Constructions at Godrevy
The local actors at Godrevy agree that the headland is an iconic coastal site in Cornwall with high natural beauty and visual attractiveness. From the perspective of human-environment interaction, the Godrevy landscape is conceptualized as a combination of appealing natural environment and human-made elements like the famous Godrevy lighthouse. In the center of this construction are the various activities that people carry out on- site and the visitors' personal attachments to Godrevy. Therefore the main management aim from this perspective is

Table 10:   Overview – The Landscape Narratives at Godrevy

Godrevy as			
	Human-Environ-ment Interaction	**Natural Systems**	**Functional Landscape**
Landscape Construction	Natural *and* human-made elements; visual attractiveness; beaches, natural beauty; the light-house, café, surf spots, walking paths, car park	Dune system, heath-land, protected sites; seal haul out; local species of national importance; Go-drevy as wildlife sanctuary; natural beauty	Godrevy as arable farmland; fertile soil, PH levels; fields used for cropping and grazing
Attachment Expressed to	Natural and built features; lighthouse as focal point; im-portance of family histories and memo-ries made at Go-drevy	The dune system, lo-cal species, flora and fauna; wild-flowers; high habitat connec-tivity; the seals	Not much attach-ment expressed; concern about ris-ing visitor num-bers and more built development
Perceived Landscape Changes	Rising visitor num-bers and connected challenges of man-aging access; pro-ceeding coastal ero-sion; built develop-ment around the headland	Rising visitor num-bers as threat to eco-systems; proceeding coastal erosion; more built develop-ment in the area	Changes in agri-cultural produc-tion through new technologies; ris-ing visitor num-bers; coastal ero-sion and cliff falls
Management Priorities	Enabling access for visitors also with bad mobility *whilst* preserving the natu-ral beauty of the Go-drevy landscape	Protection of the site and the dune system from people, cars and dogs; improve-ment of habitat con-nectivity	Effective use of Godrevy head-land for farming; effective use of new technologies for planting and harvesting

Source: interview data

to maintain access for locals and tourists to keep alive their experience with the landscape, even for visitors with bad mobility. From the natural systems perspec-tive, Godrevy is conceptualized as a nationally important dune system and coastal habitat in need of protection from harmful human activity. The priorities of landscape management here are the control of access for people to the site and the improvement of ecosystem health and habitat connectivity. From the functional landscape perspective, on the other hand, Godrevy is perceived as farmland with special requirements due to its exposed location and soils with low PH levels. The visual qualities of the landscape are not thematised in this narrative.

To summarize this subchapter and the different constructions of the Godrevy landscape, Table 10 gives an overview of the key elements of each narrative and contrasts them with each other. Drawing on the key points of these narratives, the following subchapter focuses on the normative and practical considerations that shape the relocation activities at Godrevy: the question of who should have access to the landscape, as well as the complicated local structures of land ownership and management.

6.3 Practical Considerations

As shown in chapter 6.2, with the co-existing narratives about Godrevy come different normative perspectives on how the landscape should best be managed as well as on how climate change and erosion are understood. From these different perceptions of Godrevy result distinct perspectives also on the questions of whether the headland is to be viewed as a common good and who should have physical access to it. These questions have direct practical implications for its management, especially when the coastline is eroding and infrastructure needs to be relocated. These relocation measures are additionally complicated by the local structures of ownership and landscape management at Godrevy. In the following section I present these considerations, their connections with the different landscape narratives, as well as how the local actors' concrete viewpoints shape the discussion about adaptation to coastal erosion.

6.3.1 Access to the Landscape: For whom and how?

As outlined in Chapter 3.3, when constructing a landscape as a commons, as is the case at Godrevy, a central question is that of access to this landscape. At Godrevy, this practical question of who should have access to the headland and how, is linked inseparably to the different sub-societal landscape constructions and constitutes a central point of the discussion about the relocation activities at Godrevy. When it comes to the question of access, there are two main positions among the local actors: (1) in favour of access for people and cars to the headland, and (2) critical of access for people and putting wildlife protection first. Whereas these two positions are not uncommon in conflicts around landscape management more generally (especially where a strong conservationist lobby is involved) (see 3.3), in the Godrevy case they are directly linked to the outcomes of the climate adaptation planning by the National Trust and thus constitute an important facet of comprehending the local debate around relocation from coastal erosion. The topic of access at Godrevy is divided into the perspectives of landscape as human-environment interaction (strongly represented by the National Trust) and natural systems (primarily represented by the Towans Partnership). In the following paragraphs, I will present these two contrasting perspectives on access for people to the Godrevy landscape to draw on these results in the subsequent discussion of the concrete implications for the relocation activities (6.4).

Access Yes, People in Focus

> *"There is such a long history of access here that we feel that we need to maintain that. [...] To actually just say 'Switch it off, you can no longer go there', that's a different world. We feel that's not the right thing to do. But it's a judgment call." (I-G5, 22:50)*

The National Trust describe Godrevy as "really important from a nature conservation perspective, but also from a visual perspective" and, while "within Cornwall [they] have a lot of sites, Godrevy is certainly the busiest" (I-G3, 06:02). It is emphasized that Godrevy is "very accessible, so it's a place that people can come, see the coast and really experience it", even if they are elderly or not very mobile for other reasons (I-G3, 05:09). In the center of the Trust's landscape construction is the belief that Godrevy, like many other locations, is "somewhere people come to to experience the place" to which they are attached and which they connect to their personal life stories (I-G3, I-G5). After all, I-G3 (NT) argues, "that is one of the messages of us as an organisation: we want people to come to our sites and we want them to enjoy the places that we look after" (I-G3, 50:03). Despite this wording of "looking after the landscape", the focus of landscape management here lies clearly on looking after it for the people while trying not to harm the natural environment (I-G3, 10:50). From this perspective, Godrevy is thus understood as a commons to which the Trust has the task of providing access for all groups of society (comp. 3.3, 5.4.2). Hence, denying "400.000 visitors per year" access to a site to which most of them feel emotionally strongly attached is not being considered in the decision-making process (I-G5).

Maintaining access to Godrevy is not only the wish of the National Trust, but also their legal duty under the Public Rights of Way Act (Parliament of the United Kingdom 2000) (I-G10, 23:35). Currently, I-G10 (SWCPA) "when the [coast] path is lost, the path is lost. The legal right disappears and we have to negotiate a new right of way further inland to get around that". The implementation of the new Coastal Access Act, he argues, "will help [...] to overcome that potential problem" (I-G10, 25:30) (see also 6.3.2). The British Government's Marine and Coastal Access Act 2009 he is referring to initiated the "England Coast Path"-project which requires the entire English coast to be made accessible via paths until 2020. Under this Act, if these paths are lost due to coastal erosion, the closest neighbouring inland area is given a new Pubic Right of Way automatically - simplifying matters for organisations such as the National Trust who are obliged to provide access to their properties and along the coast (I-G10, Parliament of the United Kingdom 2000, 2009). In what form they grant access to the landscape, however, and what infrastructure they provide is up to the National Trust and their local staff. Due to their strong focus on access for visitors to Godrevy, the infrastructure in form of the two car parks and the access road are therefore essential elements of the local landscape, and their pending loss through cliff erosion is an important subject of landscape management on the site (I-G3, I-G5). Maintaining access to Godrevy headland and the local beaches is not only important to the National Trust as the main land manager, but also for local businesses such as the surf school run by I-G12. Whereas at other coastal locations businesses can benefit from existing infrastructure, at Godrevy the only access and car park are those built by the Trust. Thus, local business owners are equally interested in maintaining access (I-G12, 23:35).

Although visitors are the focus of the Trust's landscape management, there is also recognition of the importance of keeping the place 'wild and natural', because after all this is what both locals and tourists appreciate about Godrevy. I-G7 (Visit Cornwall), however, underlines the importance of making visitors aware of which behaviours are harmful to the natural environment of the site (I-G7). The National Trust therefore is keenly aware of the need to protect the sensitive local wildlife, above all the colony of Grey Seals at Mutton Cove, from disturbance and pollution through visitors (I-G3, I-G5). Despite the fact that active promotion of the seal haul out could attract even more people to Godrevy, no information about the animals is shared on the National Trust website:

> "We don't actively promote it. The BBC asked us recently if they could [...] film the seals here. But we actually went back and said that at this time, we can't really cope with that and if they would mind picking another area." (I-G3, 53:09)

Hence, preserving the natural beauty of the site while "so many people come [...] is part of the challenge as land managers". Enabling people to come and "enjoy these special places" while at the same time looking after the landscape "in terms of conservation" (I-G3, 47:10). In addition to being sensitive towards the local wildlife, the National Trust are also aware of the fact that the other local actors have differing perceptions of the Godrevy landscape, even if they are not openly addressed or discussed among the different organisations in the debate around relocation. In this context, the aesthetics of the landscape are a topic that is important to many actors in the area. It is agreed among the different organisations that the infrastructure for the car park and the pathways across the headland should be visually as little impairing as possible; a requirement that the current car park – an assemblage of green fields with low hedges as dividers – theoretically fulfils (see Figure 28). When filled with cars in summer, however, it is perceived as having a strong negative impact on the natural beauty of the landscape (I-G1, I-G3, I-G4, I-G9, I-G10, I-G11) (see Figure 28). The location of the current car park on top of the headland, however, is of particular significance: Godrevy is one of the few coastal locations in Cornwall where it is possible to drive one's car directly to the cliff edge and enjoy the views across the bay (I-G1). This is important especially for people with bad mobility, as I-G5 (NT) explains:

> "Of the people who come here, half of them would be retired, elderly, not up for a two-mile-walk. A lot park down there and just sit in a deckchair outside their car to enjoy the view." (I-G5, 48:30)

Figure 28: View across current headland car park at Godrevy
Source: own photo, 2016

If that was not possible anymore, he is worried, "the vast majority of people wouldn't come. And would be upset that they can't come" (I-G5, 26:40). In addition to their own construction of the Godrevy landscape, visitor expectations thus play a large role in the management decisions of the National Trust. In this context, he argues,

> "access is important, landscape is important, wildlife is important. And what I find really interesting about this is that it is much harder managing people than wildlife really." (I-G5, 1:00:02)

When it comes to the relocation measures – framed by the National Trust clearly as adaptation to an eroding coastline accelerated by climate change – the upcoming alterations in the landscape are seen as an opportunity for change at Godrevy. As such large numbers of people do not come to Godrevy without causing problems, I-G5 (NT) sees the relocation as an opportunity to improve the traffic situation on the headland. With a thousand cars coming into the car park on busy summer days, the existing infrastructure is under great pressure and the relocation provides a chance for adjustments (I-G5, 12:20). If the car park would be gone from the headland, however, this might mean the loss of an important part of what traditionally constitutes a 'Cornish' vacation:

> "More and more sites are going to be like [Godrevy] and if you start saying in all those cases […] you only allow parking inland and a walk to the beach that would be lovely in many ways. […] But it wouldn't be that kind of tourist holiday. […] That kind of traditional English, Cornish driving a long way from home to go to a beautiful place." (I-G5, 51:10)

With this statement, he attaches meaning to the headland car park that clearly exceeds its significance on the local level: As one of the last car parks of its kind, it is a symbol for the traditional form of vacation people spend in Cornwall in which mobility by car plays a vital role.

All in all, from the HE perspective on Godrevy, and very importantly from the perspective of the National Trust as the main land manager, the local landscape is a place that people should be able to access and enjoy, no matter what degree of mobility they have. Maintaining and improving the existing pathways and infrastructure is therefore seen as the central focus of landscape management, and the question of whether the headland should be accessible by car is not negotiable. Although the visual qualities of the landscape and wildlife protection are important management tasks, they are carried out under the requirement that access to Godrevy is ensured also under changing environmental conditions.

Access No, Wildlife in Focus

Whereas it is the declared management goal of the National Trust, the AONB, the Coast Path Association and local businesses to provide access to the landscape for people of all ages and mobility, other organisations, following the NS narrative, are much more protective of the natural environment and hence critical of people and vehicles coming to Godrevy. Especially the Towans Partnership disagrees with the Trust's focus on access for the reason of wildlife disturbance in the dunes particularly, but also on the rest of the headland (I-G1). Confirming that "the whole people issue is obviously very relevant to everyone here" (I-G1, 17:20), they clearly seek to limit access to the dunes and Godrevy headland:

"We don't want to encourage too many people coming. I suppose the partnership's main thing is the conservation of the dunes, and that means the plants and the flora and fauna." (I-G1, 15:54)

At the same time, I-G1.1 admits that "it would be completely unrealistic to not be concerned with the people as well. […] People are coming here and we hope they will contribute to the economy" (I-G1, 16:10). Despite the importance of the site for both locals and visitors, he finds that landscape management must put wildlife and habitat protection first (I-G1, 16:22):

"Disturbance is a real major issue. But the National Trust are brilliant at working on that. They're just going to put some new signs on the path. They don't like signage, though, and they're quite right to. It's supposed to be a wild place." (I-G9, 37:05)

Also from the perspective of Natural England, "the pressures are quite extreme. People need space as well as nature. We need to try and get the balance right so that nature is not pushed out by people" (I-G4, 20:50). If the landscape would not be actively protected and looked after by organisations such as themselves, I-G1.1 (TP) argues, "you could have things going to wreck and ruins here just because of all those people" (I-G1, 18:40). Similarly, I-G2 finds that at Godrevy "where it's highly used by the public, it just adds an extra layer of complication" to landscape management, the landscape, in this case, being the natural environment without people in it.

For the same reason, I-G9 (CSG) "would not like [the Trust] to advertise [Godrevy more actively] unless the site is managed for increased visitors" (I-G9, 44:20). The vision of I-G4.1 (NE) in this context is that there "would be quite a lot of re-wilding of the actual coast" and that people, while they would still be allowed to visit the site, would not bring their cars to the top of the headland anymore (I-G4, 1:10:50). Also for the Towans Partnership more limited space for car parking would constitute a control mechanism for how many people can visit Godrevy at a time (I-G1). A new car parking concept is suggested by I-G4.1 (NE):

"I think there would be a possibility to leave the car park over there [at St Gothian Sands] and not have access along here. […] I don't know why they need to be able to take their cars all the way up." (I-G4, 15:20)

His suggestion would be to "make people leave their cars somewhere further away" and arrange transport by bicycles or mini buses (I-G4, 28:30). In a new infrastructure strategy, he argues, the three types of access to Godrevy headland (road, footpath, farm access) should be uncoupled so that alternative routes can be found for each purpose (I-G4, 1:12:10). When including hard-engineered foot-paths for people with bad mobility they should, however, look as natural and subtle as possible (I-G4, 1:13:10) (see Figure 29). The question of landscape aesthetics at Godrevy thereby opens up a much larger, more general discussion about whether Natural England generally welcomes people in the landscape or not:

> "If the path went up over the slope with a track [...], that is going to have a visual impact as well. [...] It's interesting, though, do we mind that? That is a really open question. Do we mind being able to see people walking along something like that?" (I-G4, 1:13:40)

Figure 29: Footpath at Godrevy with low visual impact
Source: own photo, 2016

For I-G1.1 (TP), although he does not closely connect climate change with the ongoing coastal erosion at Godrevy, the local discussion around beach access is the one issue where climate change "comes into play" in landscape management in the area. Any of the solutions the National Trust could currently implement to restore the broken steps to Godrevy Beach would, at best, be "very, very temporary" (I-G1, 45:50). At the same time, the issue of cut-off beach access is like a catalyser where "a small issue like that opened it out into what's happening to erosion across the whole of the Towans" as "that's a good sign of potential erosion for the whole area that could just collapse away" (I-G1, 13:01). A fact, he believes, that calls for the search for longer-term solutions than the reactive ones applied currently.

6.3.2 Land Management and Ownership Structures

As pointed out in the introduction to this chapter, the goal of the National Trust is to 'roll back' the existing infrastructure from the cliff edge further inland to maintain access to Godrevy headland. This approach is in line with their Shifting Shores policy, which states that one should "relocate people and buildings away from high-risk areas" (National Trust 2008: 6). Aside from the debate about whether there should be access to the landscape for people and cars or not, this roll-back is not as easily done on the ground as suggested in Shifting Shores. I-G3 points out that aside from guaranteeing access, the main management challenge when sections of the cliffs are in danger of eroding is

> "what would we do if that bit of coast path slips away? Do we own the land on the other side, is there a possibility to look at acquisition and work with landowners for future rollback?" (I-G3, 20:30)

Indeed, the structures of land ownership and management at Godrevy are complicated. Whereas the locations of the current access road and coast path are owned by the National Trust, the sections directly inland from those in are not necessarily (I-G5). According to the plans of the Trust, a section of the new road will run through the protected dune system south of Godrevy farm, which at the same time is a SSSI site. On the one hand, as pointed out in 6.1, this SSSI is managed by Natural England with the aim of protecting its flora and fauna (I-G4, I-G5). On the other hand, this land has been owned by a private landowner for the past decades who recently passed away, leaving the land in the hands of a trust fund currently managed by his children. Whereas negotiations with the former owner had reached a stage of understanding and consent, the change of ownership over the protected dunes resulted in an interruption of these negotiations with a currently unclear outcome (I-G5).

As outlined in 6.1, the local farmer cultivating Godrevy headland holds a tenancy agreement with the Trust (I-G5, I-G6, National Trust 2017b). Since a new location for the car park is likely to include areas which are currently used agriculturally, this tenancy agreement would have to be adjusted to realize the relocation measures – an adjustment that requires agreement with and compensation for the tenant farmer. The section inland from the coast path on the north side of Godrevy leading to The Knavocks, where entire sections of the path are in danger of sliding off into the sea, has already been re-routed during the past years. Further roll-back of the path, however, is currently not possible because the land in question is owned and used for grazing by a private landowner (I-G3, I-G5). St Gothian Sands, which is part of the Towans dune systems and suggested by Natural England as a possible new location for the car park, is owned by Cornwall Council and managed as a Local Wildlife Reserve by Cornwall Wildlife Trust (I-G2, I-G4, I-G5). Being a protected site, it therefore cannot be converted to a different purpose easily (I-G5).

Both the contents of the Trust's 'Shifting Shores' strategy (National Trust 2008) as well as the perspectives of the local actors at Godrevy suggest working with natural processes when it comes to coastal erosion. However, simply rolling back the access road and car park further inland shows to be a very ambitious goal in practical terms. Whereas theoretically there are a number of locations available at Godrevy, the relocation activities are complicated by the local structures of land ownership and management. A braid of National Trust property, tenanted farmland, privately owned land, as well as protected wildlife sites significantly limit the options the National Trust has for moving the infrastructure further inland to maintain access to the headland (see Figure 30).

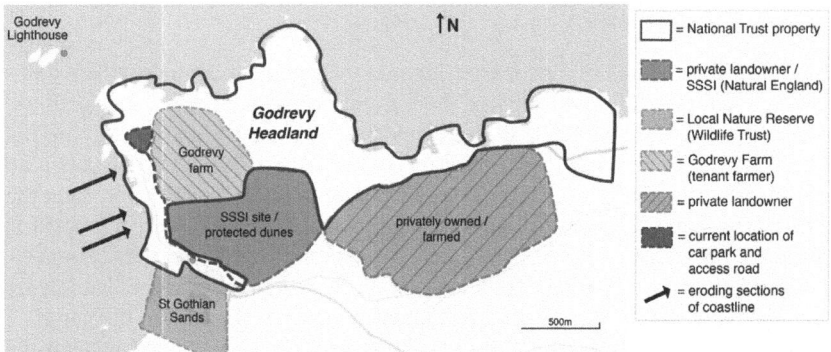

Figure 30: Ownership Structures at Godrevy
Source: © OpenStreetMap Contributors (www.openstreetmap.org), modified; interview data

Summarizing the Practical Considerations about Relocation
Landscape change through accelerated erosion, here framed clearly as impact of climate change by some actors, sparks a re-negotiation of what the Godrevy landscape is, and for whom it is managed. Whereas the landscape constructions and the different organisational guidelines and policies suggest roll-back from the cliff edge and the support of natural processes, the interviews with local actors show that when it comes to practical action a number of issues must be complied with that exceed these normative considerations. These issues are on the one hand the practical question of who should have access to the headland and how, and on the other the structures of ownership and land management in the area. The National Trust thereby directs its management towards enabling access to the landscape and the coast path for all visitors with or without their cars. Their slogan "Forever, for everyone" thereby plays a huge role at Godrevy, making it their main task to maintain access to the landscape. Having a car park and access for large numbers of people on top of the headland thus is not negotiable, even under a changing climate and challenging conditions for landscape management (I-G3, I-G5). Other local actors focus on the local wildlife and protection of its habitats, a focus resulting in a very critical view of access to the site for too many people. Moreover, the ownership structures constitute barriers to adaptation that relativize the ambition of the local actors to relocate the access road and coast

path further inland or, if solutions are negotiable, at least significantly lengthen the process of reaching agreement about new locations for the infrastructure on the headland. These practical considerations thus also have concrete implications for the discussions about the physical-material reshaping of the Godrevy landscape through the relocation of the car park, the coast path and the access road, as I show in the following section.

6.4 Mapping it out: The Different Perspectives on Physical-Material Relocation at Godrevy

From both the co-existing constructions of the local landscape (cf. section 6.2) and from the practical considerations about adaptation to coastal erosion (cf. section 6.3) result concrete implications for the physical-material relocation activities at Godrevy. The interviews showed that the contrasting conceptualizations of the landscape, management priorities, and understandings of climate change and coastal erosion are central to comprehending the viewpoints of the different actors involved in the process.

Moreover, the discussions about access and the ownership structures influence the decision-making process in practical terms. It is the interplay of these factors which results in five concrete, differing viewpoints among the local actors on how Godrevy should be reshaped physically to adapt it to coastal erosion. In this section, based on the underlying landscape constructions and the practical considerations laid out in 6.2.1.-6.3.2, I present these viewpoints accompanied by maps illustrating the different actors' preferred locations for the new car park and access road (see Figures 31– 35). For reference and comparison with these illustrations, see Figure 18 (p. 138) which illustrates the current position of the headland car park and the existing access road.

Suggestion 1: Conversion of Farmed Fields
For those actors who perceive the Godrevy landscape in terms of human-environment interaction, above all the National Trust, the main priority in the relocation process is to maintain access to the headland for people and cars. Nonetheless, the actors set emphases on different characteristics of the landscape, resulting in varying preferable locations for the new infrastructure. It is primarily the National Trust, the AONB Partnership, and the owner of the local surf school who have concrete suggestions for where, and where not, to build the new road and car park. The National Trust as the main land manager of Godrevy wants to proactively approach coastal erosion *before* the road erodes into the Atlantic. In doing so, they seek to ensure that all visitors, with good or bad mobility, can continue to enjoy the special place that Godrevy is from their perspective (I-G3,

I-G5, I-G11). Their temporal planning scope is twenty to fifty years into the future, which in practical terms means that the new road will be located so far inland that it lies safely behind the estimated erosion line of in fifty years' time. Access and practicability for visitors thereby stand in the focus, especially for the elderly and those with bad mobility. Proposals to withdraw parking and car access from the headland and let nature take over, as I-G5 (NT) argues, "would make [his] job really, really easy!" (I-G5, 22:14). Due to their perception of the Godrevy landscape and their position in the debate around access, however, it is *not* an option for the Trust to relocate the car park entirely away from the headland (I-G3, I-G5).

Whereas the National Trust seeks debate with Natural England and the AONB Partnership for legal reasons as these organisations have veto rights in the planning process, the local staff also consider it important to find consent among all other involved actors such as the tenant farmer to "work alongside our neighbours, so that we can have a broader view than just managing our own land" (I-G3, 12:45). Moreover, I-G5 (NT) explains, "there are very strong opinions [...] from people who visit. For lots of people this is their favourite place in the world, and just to say 'Oh sorry, you can't go there'" would cause resentment among both tourists and the local population (I-G5, 26:40). Very aware of divergent perspectives on the relocation activities at Godrevy, he understands that all these actors are focusing in their own best interests. Although the National Trust owns Godrevy headland, he considers an open, transparent and participatory process as very important in this context. While keeping the focus on access to the landscape, he stays flexible with regards to the specifics of the relocation plans:

> "If people who are very experienced come along and say 'Actually there is a better solution here' or 'We really think this is the wrong choice', or if every single person I spoke to was saying this is a really bad idea, I might think that perhaps I got this wrong!" (I-G5, 59:05)

During the empirical phase in early summer 2016, discussions about the car park and road relocation were still ongoing. The Trust, however, approaches coastal adaptation with a great sense of urgency and has already mapped out their preferred solution. Figure 31 illustrates the concrete location that the National Trust suggests for the new car park and access road. Their current plans consist of the following activities: Removing the current headland car park, re-wilding the area and creating a habitat of coastal protected heathland; reducing the current land area tenanted by the farmer by three fields, converting the largest one into a new car park and re-wilding the other two; as well as re-routing the access road and coast path through the protected dunes south of the farmed fields and away from the cliff edge.

The Trust is currently negotiating with the private land owner in possession of the protected dunes north of St Gothian Sands (see Figure 31) to acquire a section of those dunes; an endeavor complicated by the passing away of the original owner of the site in early 2016 and a currently unsettled ownership situation (I-G5, I-G8) (see 6.3.2). Whereas in an "ideal world where you owned everything and you had unlimited money" the National Trust could decide freely where to relocate their infrastructure to, existing property structures and landscape designations decisively limit the origination's scope of action (I-G5, 25:10). The tenant farmer moreover, who would no longer be able to use three of his fields for agriculture when the car park gets relocated, would be compensated financially (I-G5). This conversion of farmland is seen by the National Trust as a solution with few disadvantages as "moving the car park into the farmland over there which has little wildlife value […] will be a value to the landscape and to wildlife as well" (I-G5, 19:45).

I-G12, owner of the local surf school, also perceives the Godrevy landscape in terms of human-environment interaction. As his business depends on visitors at Godrevy, he is much in favour of the Trust's relocation plans and welcomes access to the headland by car. "If they lose the access here" he argues, "that's a lot of parking gone! A lot of tourists that can't come here not having that infrastructure" (I-G12, 22:50). Regarding the time scale of the relocation, he

> "reckon[s] it'll happen, I think it's pretty close. I don't think I get told as much as I would like to be told, but I think they're getting close - which is going to be a great thing for me." (I-G12, 22:50)

Although generally against having a car park on top of Godrevy headland, the representatives of the Towans Partnership surprisingly support the Trust's suggestion for relocation. Their main motivations for this standpoint are both "visual and practical reasons" (I-G1, 23:40). Not concerned about Natural England's SSSI site, they welcome the plans of the Trust to re-route the road and use some of the farmed fields for parking (see Figure 31). Not only speaking for themselves, I-G1.1 recounts that

> "certainly at the Partnership everyone is quite happy with the proposal [by the National Trust], because they are going to move the car park away from the edge of the sea – even if through some of this land – and through the agricultural land that they've got." (I-G1, 20:45)

Concerned with more visitors coming also to the section of land they are responsible for, the Towans dune system, the interviewees from the Partnership argue for replacing the old car park with one that has is smaller than the current one. "That's always the control you've got, it's where people can park", I-G1.1 argues (I-G1, 18:58). In the Towans, he explains,

"There are no places where cars are able to access the dunes and drive out. [...]
Because it does cause a huge impact! Motorbikes sometimes did go through and you
do get some impacts of that. But cars - that's not happening!" (I-G1, 23:10)

Thus whereas the National Trust to a certain degree subordinates the protection
of wildlife and historic features to enabling access to the landscape for visitors,
the Towans Partnership are not willing to make this compromise – to the disad-
vantage of people with bad mobility not being able to move easily across the
dunes.

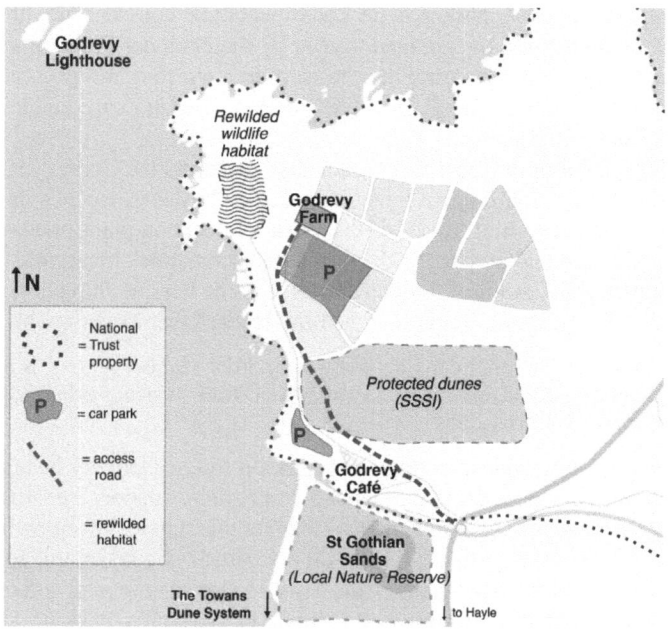

Figure 31: Map: Suggestion 1 – Conversion of Farmed Fields
Source: © OpenStreetMap Contributors (www.openstreetmap.org), modified interview data

Suggestion 2: Field Parking with Low Visual Impact

Also seeking to maintain access to the landscape for people, the AONB Partnership at the same is concerned about the visual impacts of any newly-built infrastructure at Godrevy. Focused on preserving the local distinctiveness of building
styles and overall landscape character, for I-G11 (AONB) the most pressing
questions in this context are:

"Can you see the road from up here, would it have an impact on the setting? Would it have an urbanising effect? How many people would experience that as a change in the landscape? And would it be sensitive to that? Would it change the character of the place?" (I-G11, 33:20)

Therefore, landscape management for her is ultimately about "finding ways of doing that which are compatible with the aims of landscape conservation, which are the aims of the National Trust and the aims of the AONB" (I-G11, 32:20). Godrevy, she argues, is the first example in Cornwall in the context of adaptation to climate change and coastal erosion where these goals do not conform (I-G11). She is especially critical of the Trust's tendency to give more weight to practical considerations than to protecting the character of the landscape:

"Everybody speaks about landscape in a positive way and recognizes its beauty, and people like the National Trust, you know, it's their core business to restore landscape and preserve it – until they earn money from it, and then…" (I-G18: 13:02)

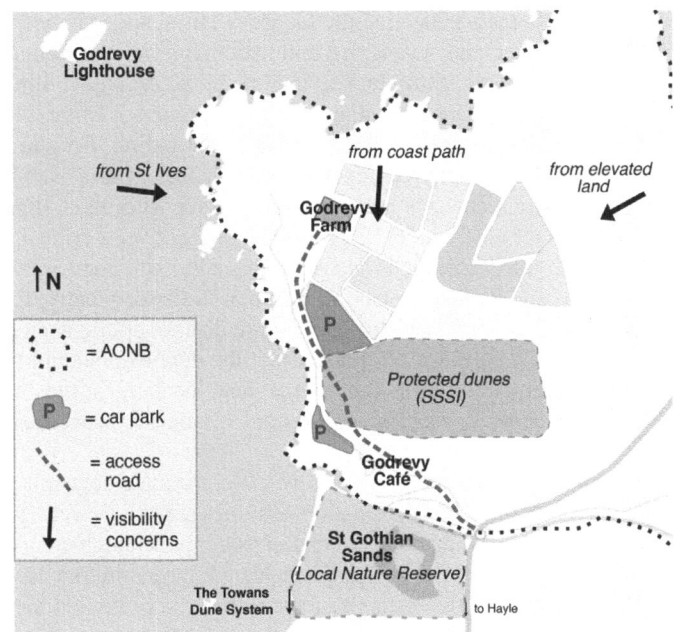

Figure 32: Map: Suggestion 2 – Field Parking with Low Visual Impact
Source: © OpenStreetMap Contributors (www.openstreetmap.org), modified; interview data

Pointing out the need for access to the protected landscape, she hopes that in the future the Trust "will be able to put something in like a disabled route so that people with bad mobility will still be able to get up here and enjoy it" (I-G11, 28:00). A special concern for the AONB Partnership, however, is the potential visibility of these new structures in the landscape from across the bay in St Ives, from different points of the coast path, as well as from parts of the headland with higher elevation. Consequently, while agreeing with the National Trust that the currently farmed fields are a suitable position for the new car park, it should be located in the field closest to the cliff edge to be hidden from sight behind the hills (see Figure 32). In their new management plan, the AONB Partnership

> "supports the restoration of this area and the removal of the road. [...] [Landscape] character, wildlife, just to restore it to a more natural habitat is what we'd actually like to see out here. This is a nice headland, it shouldn't really be covered in cars." (I-G11, 23:00)

Suggestion 3: Conversion of St Gothian Sands
From the natural systems perspective the Godrevy landscape is conceptualized in terms wildlife and habitats, a sensitive and nationally important dune system which is under threat from harmful human interventions. Active wildlife protection stands in focus of the landscape management approach of Natural England, the Towans Partnership, as well as the Cornwall Seal Group. Coastal erosion, although not necessarily seen in relation to climate change, is perceived as a positive development which changes habitats, and the support of natural processes is viewed as important here (see 6.2). From this perspective, a critical view on access for people to the landscape is derived, especially with dogs or by car, for reasons of wildlife and habitat protection (see 6.3.1). Consequently, the actors following the natural systems narrative have quite distinct opinions on the relocation activities. While agreeing on the fact that the new road should not cross any protected sites, however, their ideas of the ideal location for the new structures differ widely depending on which section of land they specifically look after.

As an organisation with focus on local wildlife, the main concerns of Natural England when it comes to the relocation activities are (a) protecting the sensitive habitats on the headland and to (b) making sure that enough compensation areas will be created if some of the current fields are turned in to a car park (I-G4, 08:51). Although the Trust offers to restore the area of the old car park to coastal heathland, for the interviewees from Natural England

> "there's lots of reasons actually to think why we shouldn't encourage a new car park at this location or the realignment of the road. Rather let the land here revert to dune grassland, maritime grassland." (I-G4, 58:55)

Their major worry in this context is that the route for the new road suggested by the National Trust would cut through the protected dune system south of the farm which is a SSSI, a Site of Special Scientific Interest cared for by Natural England (see Figure 33). "Sites of Special Scientific Interest (SSSIs) are protected by law to conserve their wildlife or geology. […] There are certain things you can't do on SSSI land without Natural England's consent" such as, for example, building a road across the site (Natural England 2013). Due to this statutory protection "it's not easy to re-draw the line of the SSSI" (I-G4, 8:58), and Natural England must give consent to the National Trust for doing this. Generally questioning the need for a car park on the headland, however, Natural England is skeptical towards access for people to the landscape and wonder "why does there need to be a car park here anywhere?" (I-G4, 11:55). Planning with a time horizon of several decades and considering the projected rate of erosion, I-G4.1 thereby stresses the need for a long-term solution that preserves both the wildlife and historic features on the headland (I-G4, 07:30). Moreover, a new car park should not disturb the visual, 'wild' landscape character of Godrevy headland or the surrounding area (I-G4, 57:05). Re-wilding the headland would thus have two benefits; I-G4.1 (NE) underlines: the obvious gain in habitat connectivity and wildlife value on the one hand, and the mitigation of the visual disturbance through cars reflecting the sun on the other (I-G4, 11:20).

Considering all these concerns, the interviewees from Natural England propose relocating current car park entirely away from the headland and placing it onto St Gothian Sands, a former sand mining pit south of Godrevy, instead (see Figure 33). Overlooking St Gothian Sands from an observation deck, I-G4.1 (NE) argues:

> "We've got this area here. That's the old sand works there, St Gothian. [...] Now it's a Local Nature Reserve. I mean, the visual impact there would be far less with [the car park] being here and it wouldn't touch any protected land. So there would be no losses." (I-G4, 13:10)

Surprisingly, the Natural England staff suggest relocation to St Gothian Sands despite of the fact that it, too, is a protected wildlife site. In contrast to the SSSI site in the dunes which Natural England is responsible for, however, this site is looked after by the Cornwall Wildlife Trust. While the protection of local ildlife is an important aim of Natural England, there nonetheless seems to be a hierarchy of which types of wildlife are more important to look after than others. Aware of this suggestion by Natural England, I-G5 (NT) also admits that he "was a bit taken up by [them proposing] that, because that's a wild, natural area in its own right" (I-G5, 24:15). A relocation of the car park and road to St Gothian Sands as suggested by Natural England is therefore not considered by the Trust for two reasons: on the one hand it is a protected habitat looked after by the Wildlife Trust, and on the other it is owned by Cornwall Council and not currently possible for the National Trust to acquire (I-G5).

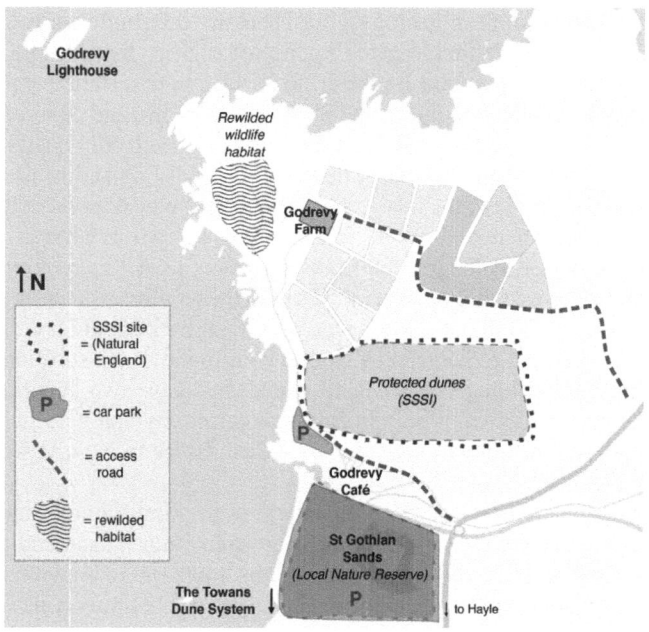

Figure 33: Map: Suggestion 3 – Conversion of St Gothian Sands
Source: © OpenStreetMap Contributors (www.openstreetmap.org), modified; interview data

Suggestion 4: Subterranean Visitor Center
Although following the natural systems narrative, I-G9 (CSG) supports access by car to Godrevy headland. This acceptance is partly related to her personal experience with the place:

> "Lots of people, like my mother, will happily come here and sit in the car and enjoy the view, but she won't get out. And if you're putting parking in a field here, that may be less possible." (I-G9, 1:04:00)

The main concern of the Seal Group is the colony of Grey Atlantic Seals which haul out in Mutton Cove on the north side of Godrevy headland. As the path from the current car park to the sensitive wildlife site of Mutton Cove is very short and leads to significant disturbance of the animals, I-G9 would appreciate the new car park being further away from the site she looks after. A longer walkway would give the National Trust the opportunity to install signage alerting the visitors to be quiet and would separate the noisy car park geographically from the seal haul out (I-G9, 42:20). After all, she argues, "it's not a small amount of parking we're talking about here! It's a massive amount of parking" (I-G9, 1:07:00).

When altering the Godrevy landscape to build a new car park and access road anyway, I-G9 suggests that the visitor experience and wildlife research on the site could be strongly improved by locating a subterranean information center about the local seals, dolphins and other species near the current car park (see Figure 34). To not disturb the landscape visually, however, "it would be great to have something under the dunes" (I-G9, 48:35). Considering the entire relocation process, I-G9 agrees with the Trust in her sense of urgency regarding the proceeding of coastal erosion on the site:

"I have a feeling that this road is going to be closed unexpectedly one day because there has been a pitch point fall and they just had to close it. And at the point they'll be forced to think about building another one." (I-G9, 55:20)

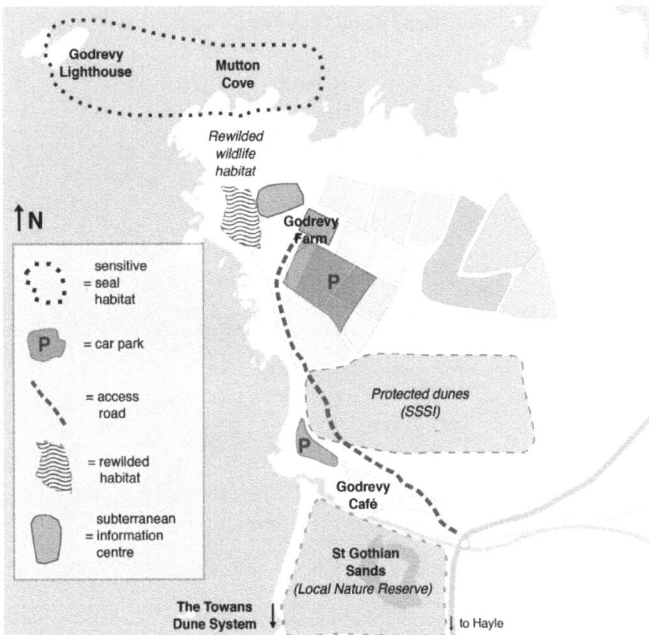

Figure 34: Map: Suggestion 4 – Subterranean Visitor Center
Source: © OpenStreetMap Contributors (www.openstreetmap.org), modified; interview data

Suggestion 5: Relocating Road, but not Car Park

Whereas the Trust and the Towans Partnership agree that the farm land they wish to convert is not a big loss regarding land use and biodiversity, the tenant farmer at Godrevy perceives the land he cultivates very differently. For him, his fields and the Cornish hedges between them are a haven for wildlife (I-G6, 21:20). Instead of sacrificing three of his fields, he prefers the car park to stay in the current location and attaches little value to the maritime heathland which Natural England and the Wildlife Trust view as important habitat (I-G6, 18:40) (see Figure 35). Moreover, he argues, having a car park in front of the farmhouse of Godrevy Farm will impact negatively on the property's value – a fact which is especially important as he, himself, recently moved out of the farmhouse which will either be converted by the National Trust into holiday homes or rented out as a private home (I-G6, 36:35). Although he has his own specific perspective on the relocation activities at Godrevy, however, the tenant farmer understands well that many different interests must be taken into account in the relocation process:

> "I don't want to lose three fields. I don't! But the Trust look at it in a different way and they got to deal with all the people coming up here." (I-G6, 36:35)

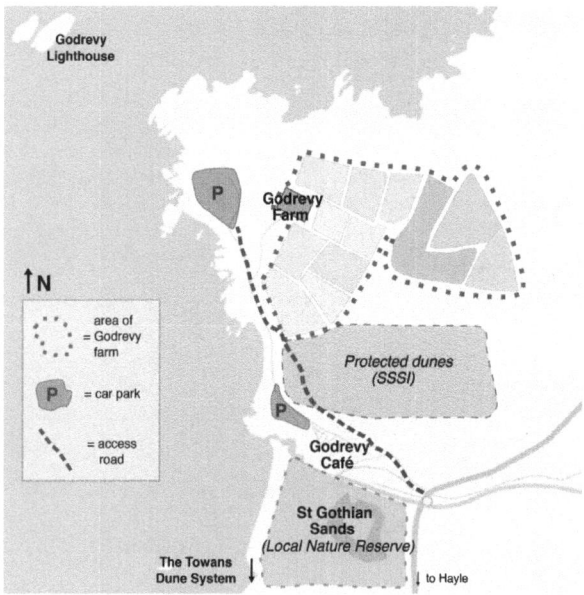

Figure 35: Map: Suggestion 5 – Relocating Road, but not Car Park
Source: © OpenStreetMap Contributors (www.openstreetmap.org), modified; interview data

6.5 Preliminary Conclusions

Although the adaptation measures planned at Godrevy have a small spatial scale, they constitute a valuable in-depth case study that highlights the influence of differing conceptualizations of a landscape for its reshaping in times of a changing climate. The landscape narratives in their local context result in distinct approaches to landscape management and perceptions of climate change and coastal erosion, all of which underpin the local actors' viewpoints on the relocation activities. Among others the National Trust and the AONB Partnership construct the Godrevy landscape in terms of human-environment interaction and focus their efforts on maintaining access to the headland, also in times of changing environmental conditions. For Natural England, the Wildlife Trust and the Towans Partnership, on the other hand, Godrevy headland is a sensitive wildlife habitat of national significance that must be protected from human activities. The tenant farmer, moreover, views the headland as a space for agricultural production that should be used as effectively as possible without disturbing the balance of the natural environment. Hence, the local actors following these narratives lay the foci of their management activities on different facets of the Godrevy landscape. At the same time, they perceive climate change and coastal erosion differently. From the HE perspective, climate change is accelerated by human activity and connected to the rapidly eroding coastline, resulting in a perceived urgency regarding adaptation to this erosion. The actors following the natural systems narrative, in contrast, understand climate change and coastal erosion rather as parts of a natural circle and perceive the erosion to positively induce habitat change in the area.

In addition to the normative considerations of the different landscape narratives and perceptions of climate change and coastal erosion, the relocation activities at Godrevy are influenced by practical considerations and limitations. The staff of the National Trust thereby take on a key role as facilitators of the discussion process and are seeking to actively involve the other local actors. While the National Trust's Shifting Shores strategy is a "high level strategic document focused on ends and not means" of adapting to a changing coastline (GEOGHEGAN & LEYSHON 2014: 640), the ownership and land management structures at the local level limit the options of the National Trust of where to place the new car park and road. Moreover, the question of who *should* have access to the landscape and in which form is controversial among the different organisations at Godrevy; whereas the Trust wishes to provide access to the headland by foot and car based on their organisational policies and conceptualization of the landscape, other actors such as Natural England and the Towans Partnership are highly critical of visitors having access to the sensitive local habitats.

Both the normative considerations about the Godrevy landscape and the differing perceptions of coastal erosion lead to distinct, mappable viewpoints on the ideal locations for the new road and car park. However, even though some of the local organisations share certain landscape narratives (e.g. Natural England and Towans Partnership), these organisations often prefer locations that are far apart from each other. The Godrevy case study is thus a valuable example of landscape management in the face of climate and coastal change at a location where various landscape designations overlap and solutions must be found among a number of actors with different perspectives on the local landscape.

After having presented the empirical results of this study, I discuss these findings in the next chapter by linking them back to the research questions and theoretical framework. I thereby consider their relevance in the context of constructivist landscape theory and their implications for understanding climate adaptation activities on the local level as multi-actor processes in their societal embeddedness. After this discussion of the empirical results, a critical reflection will follow of the theoretical background of this study, the methodology applied, as well as the role of the researcher in in-depth qualitative studies such as this one.

7 Discussion of Results

It becomes clear from the empirical findings of this study presented in chapters 5 and 6 that the same physical-material spaces, in this case Cornwall and Godrevy, can be superimposed with very contrasting constructions of the local landscapes by different actors in a society. Understood as particular forms of people-place relationships, these landscape constructions show commonalities as well as differences regarding what the landscape is, and how it can best be adapted to the impacts of climate change. The three dimensions of landscape construction proposed by Kühne (2018) are mirrored in the accounts of local landscapes in the empirical data of this study: 'societal landscape constructions' shared between the interviewees (cf. sections 5.1, 6, 7.1); 'sub-societal' constructions in form of the co-existing landscape narratives among the local actors (cf. sections 5., 6.2); and 'individually updated' constructions by individual persons working for local landscape management organizations (cf. sections 3.4.1, 5.3.5, 6.2). I discuss in this chapter how these landscape constructions and people-place relationships on the collective, sub-societal, and individual level shape specific perspectives on climate change and adaptation in Cornwall and Godrevy (7.1 – 7.4). At the end of this chapter, I draw on these findings to develop a place-based framework for local climate adaptation, synthesizing what facets of people-place relationships influence how adaptation to climate change is carried out in its local and societal context (7.4.3). Lastly, I critically reflect the theoretical approach of this study, its methodology and research design, my own role as a researcher in the process, as well as the limitations of this study (7.5).

7.1 Climate Adaptation, Place Attachment, and Regional Identity

Fresque-Baxter & Armitage (2012: 253) argue that it is vital to move people-place relationships into focus when investigating the role of subjective values on adaptation to climate change. A worthwhile starting point in this context are the shared attachments that the interviewees of this study have to Cornwall as a place and its landscapes. I showed in the previous two chapters that the landscape narratives about Cornwall and Godrevy and the interviewees' personal accounts of the Cornish landscape reveal high degrees of place attachment (cf. sections 5.3, 6.2). Attachments exist to Cornwall's visual qualities with the landscape described as "a jewellery box that is ridiculously beautiful" (I-7: 4), to its natural environment in general, but also to its local distinctiveness. Such statements

© Springer Fachmedien Wiesbaden GmbH, part of Springer Nature 2019
V. Köpsel, *New Spaces for Climate Change*, RaumFragen: Stadt – Region – Landschaft, https://doi.org/10.1007/978-3-658-23313-6_7

underline the importance of the topic of landscape for those living in Cornwall, and show that the interviewees are personally attached to its landscapes; a fact that influences also their decision-making about climate adaptation (cf. section 7.3). Nonetheless, the question remains if strong attachments to places and landscapes hinder or support the acceptance of change through climate adaptation measures (cf. Devine-Wright 2014). A closer look at some facets of Cornish regional identity gives insight into how such processes of climate adaptation are embedded in local contexts and shaped by shared constructions of place and landscape.

Climate Adaptation, Shared Landscape Perceptions, and Regional Identity
As outlined in section 1.3.1, landscape constructions from existing literature in Cornwall focus on its mining and industrial heritage; beautiful scenery; mythology and Celticness; fishing, farming, and rurality; tourism and surfing (cf. Deacon 2010, Tregidga 2012). These topics feature also in the shared constructions by the interviewees. The visual characteristics of Cornwall as an attractive coastal region thereby consistently play an important role in their accounts. Prevalent local debates about the landscape, the image drawn by the tourism industry and popular media, and current reproductions of the Cornish past presumably have an important influence on these constructions (cf. section 1.3.1). As the official policy narrative clearly shows, these shared perceptions of the landscape have decisive weight in processes of planning, regional development and marketing, and the tendency to romanticise Cornwall's past (cf. section 5.2).

Three facets of Cornwall's regional identity influence adaptation activities on the peninsula: a strong notion of conservatism which is in tension with a spirit of innovation; an 'end of the land'-mentality among the Cornish population (cf. sections 5.1, 5.4.1); as well as a sense of uniqueness of Cornwall as a place. Firstly, there is a strong tendency in Cornwall to reminisce about the past and hold on to existing structures. In the context of climate adaptation, this "innate conservatism" (I-3, Par'lor) implies a strong focus on preservation of the status quo of Cornwall's landscapes at the expense of the implementation of forward-looking, innovative adaptive measures. Physical-material landscape elements in focus of such preservative activities in Cornwall are, for instance, the relicts from the mining era (cf. section 5.3.3). Devine-Wright (2014: 172) argues in this context that both the impacts of climate change and human responses to it may disrupt people's sense of continuity with a region's past (cf. section 2.2); a fear that becomes apparent also in the attempts to preserve the status quo of the Cornish landscapes. In this sense, the question by ibid (2014) if strong attachment to place is a barrier to adaptation must be answered positively; if this attachment leads to the conservation of existing structures even if that means to risk their damage through flooding or coastal erosion, it constitutes a barrier to successful adaptation to the changing climate. Opposed to this conservative facet of Cornwall's regional identity stands a spirit of innovation also rooted in the region's mining and industrial history. As one of the forerunners of British industrialisation, Cornwall's past miners and engineers understood themselves to be at the forefront of innovation and invention; a mentality that is still shared today by parts of the county's population. However, in Cornwall's current political situation this innovative thinking is superimposed by the conservatism that, according to a number of interviewees, dominates also the decision-making within Cornwall's Council (cf. section 5.4.1). The most noticeable example for this dichotomy is the debate about renewable energy infrastructure in the region (cf. sections 5.1, 5.2, 5.3.3), but also the debates about responding to climate change in the form of either engineered structures or natural solutions (cf. section 5.4.2). This dominance of a conservative, backwards-looking spirit within the Council is also mirrored in the absence of an overarching and cross-sectoral adaptation strategy in Cornwall (cf. sections 1.3.2, 5.4.1).

The third facet of regional identity with relevance for adaptation processes is the 'end of the land'-mentality among Cornwall's inhabitants (cf. sections 5.1, 5.4.1). Due to its geographical location and relative remoteness from London as the political center of England, the Cornish population has long approached challenges with a spirit of self-reliance and independence from political decisions taken elsewhere. On the one hand, as I-6 (Counc) points out, this mentality results in a tendency of "building resilience into our communities" and aiming for self-sufficiency also when it comes to adapting Cornwall to the changing climate (cf. 5.1, 5.4.1). On the other, this rejection to be superimposed with solutions decided elsewhere underlines yet again the importance of considering place-values and societal constructions of places and landscapes in Cornwall when it comes to the implementation of adaptation strategies on the ground; both from the perspective of what shapes local adaptation processes, but also with regards to the acceptance of adaptation measures by local actors and the local population.

Lastly, what influences adaptation activities in Cornwall on the collective level is a sense of uniqueness of the region and its landscapes shared among local actors (cf. section 5.1). The perception of Cornwall as a place distinct from all other parts of the UK thereby constitutes both a barrier and a motivator for active climate adaptation. On the one hand it creates high incentive to take up action and protect Cornwall' landscapes from harmful climate change impacts. Depending on the underlying landscape narrative and perception of climate change and – in the case of Godrevy – coastal erosion, however, this ethos of protection can result either in action or intentional inaction regarding a response to climate change (cf. sections 7.2, 7.4).

The Symbolic Significance of Godrevy for Climate Adaptation in Cornwall

Regarding the societal level of landscape construction in Cornwall, the case study of Godrevy provides interesting insights into how a particular place can hold symbolic significance in the context of climate change and adaptation. Godrevy headland emerged as an in-depth case study from the interviews of the first empirical phase, and is described by the interviews as a case with particular symbolic significance for societal processes of climate adaptation in Cornwall for a number of reasons. The headland is perceived as a visitor destination of national importance, is one of the most-visited sections of coastline in Cornwall, and the only one where the coast is directly accessible by car. The vast offer of outdoor activities for visitors of all ages as well as its natural beauty make it especially attractive for both locals and tourists, drawing hundreds of thousands of people to the site yearly and supporting the local economy. Regarding its landscapes, Godrevy is collectively perceived as visually attractive, iconic, and as representative for the beauty of Cornwall's coastlines (cf. section 6, introduction). Moreover, the site has particular significance in the eyes of the National Trust with regards to the growing part of the population with bad mobility (cf. 6.3.1, 7.4.2). In addition, the high significance Godrevy is given in its regional context results in high vested interests on the sides of all local actors in protecting what they perceive of as its landscape.

More importantly in the context of this study, however, is the symbolic significance of Godrevy for conflicts around adaptation to climate change. Godrevy is described by the local actors as the first example in Cornwall where different landscape management organisations, ones that usually work harmoniously alongside each other, disagree about the ideal solution for adapting to coastal erosion (cf. section 6, introduction). Especially the larger organisations such as the National Trust, the AONB and Natural England are interested in finding a joint vision for Godrevy; so far, however, without success. The changing environmental conditions, differing priorities, as well as the different narratives about the Godrevy landscape thereby constitute catalysers for conflict – a conflict that revolves primarily around the question of access to the landscape, and that can only be understood in its local context and by uncovering the underlying values and place constructions of the local actors (cf. section 7.4).

I discussed throughout this section that adaptation activities in Cornwall on the societal level are influenced by local actors' strong attachments to Cornwall as a place, and a high importance of the phenomenon of landscape in Cornish regional identity. Moreover, particular places such as Godrevy have symbolic significance for local discussions about climate adaptation, and can be viewed as heralds for future conflicts between local actors. In this regard, this study's findings are line with the arguments by Deacon (2010) and Tregidga (2012) who conclude that the topics of landscape, place, and the natural environment importantly shape societal processes in Cornwall. The empirical data supports also the argument by Gailing (2012a) that societally shared landscape constructions provide a basis for "regional [...] individual and collective identities, regional utopia [...], [as well as] regional ideologies" (ibid: 1989). However, I argue in this study that shared place attachments and a strong regional identity alone do *not* result in consensus about how to adapt local places and landscapes to a changing climate. The crucial point here, I find, is not the binary question of "Are people attached to this place or not?". Much more importantly, we need to ask *what* place and *what* landscape exactly local actors have in mind when looking at a particular physical stretch of land, and how these perceptions influence their decisions about climate adaptation. The landscape narratives identified from the interviews in Cornwall and Godrevy provide important answers to these questions.

7.2 Distinct Adaptation Approaches Grounded in the Landscape

It was the aim of this study to investigate if and how different approaches to climate change adaptation are influenced by different actors' perceptions of the local landscapes in which they live and with which they work. This study shows that the answer to the first part of this question clearly is yes. Delving deeper under the surface of the shared perceptions of Cornwall's landscapes, quite contrasting landscape constructions emerge with perhaps surprising differences to each other. Those narratives shared only between some of the actors on the regional and local level represent distinct sub-societal landscape constructions as conceptualised by Kühne (2008, 2018) (cf. section 3.4.1). They are neither absolutely distinguishable from each other nor are they mutually exclusive. Nonetheless, they reveal important insights into local actors' reasoning behind their decisions around landscape management and climate change adaptation. The empirical findings of this study bring to light five distinct narratives about the Cornish landscapes: alongside the official narrative of the Council (cf. 5.2) stand societal constructions of the landscape as human-environment interaction, natural systems, visual beauty, as well as functional spaces of production (cf. 5.3). When comparing these narratives, there are five key points in which they differ from each other and that influence local perspectives on climate change adaptation: the elements that are viewed as constitutive for the landscape and distinct interpretations of the same; landscape management priorities and the question of *what* should be preserved; understandings of nature and the human-nature relationship; as well as perceptions of climate change and its impacts. As outlined

Table 11: Overview on Landscape Narratives and their Symbolic, Aesthetic, Cognitive, Emotional, and Normative Dimenions (part I)

Landscape Narrative	Constitutive Elements	Aesthetic Value	Landscape Knowledge
Human-Environment Interaction	Natural human-made elements; expressions of human dwelling; settlements and local human practice	Natural beauty; local distinctiveness of building styles; mining heritage; Cornish hedges and fields; coastline and beaches	Historic and current human-made landscape features; landscape genesis; effects of human dwelling; sustainable management
Natural Systems	Local wildlife and habitats; rare species; flora and fauna; agricultural land; rivers, coastline, heathland, dunes	Vegetation and natural elements; 'pristine' habitats; ocean and coastline if untilled; rolling hills; green colour	Ecosystem functions; local fauna and its behaviour; local vegetation; growing seasons; ecology; biodiversity
Visual Beauty	Natural and historic human-made elements; natural environment; green-looking countryside; coast; beaches	Local distinctiveness in form of beautiful scenery, light; traditional building styles; views across fields and ocean	Little knowledge about the different landscape elements, their history, or ecology
Functional Landscape	Industrialised country-side; human-made and hardly natural; agricultural space; used for food and energy production	Not of relevance; engineered structures are not objected if contributing to efficient use of landscape	Agriculture and local ecosystems; efficient land use; renewable energy infrastructure; environmental/economic sustainability

Table 11: Overview on Landscape Narratives and their Symbolic, Aesthetic, Cognitive, Emotional, and Normative Dimenions (part II)

Landscape Narrative	Attachments	Foci of Preservation	Construction(s) from Literature
Human-Environment Interaction	Attractive natural elements; historic features; locally-run businesses; villages; landscape as expression of identity	Local distinctiveness in building; local human practice; sustainable livelihoods and communities	*Landscape as home; landscape as recreation; landscape as memory*
Natural Systems	Distinctiveness of local species; environmental sustainability; healthy nature environment; green spaces	Local distinctiveness of species and habitats; sensitive ecosystems; flora and fauna; habitat connectivity	*Landscape as natural; landscape as rural; landscape as object*
Visual Beauty	Natural beauty; historic features; attractive buildings like manors or cathedral; green colour of landscape; light conditions	Visual features of the landscape; natural looks; local buildings styles; appearance of natural and cultural heritage	*Landscape as natural; landscape as beautiful and good; landscape as visual*
Functional Landscape	No strong attachment expressed except for Cornwall as a place of residence	Preservation not in focus; efficiency of landscape management prioritised; built structures not viewed critically	*Landscape as rural; landscape as an object*

in section 5.4.2, from these points result four distinct views on climate change adaptation in Cornwall (cf. Köpsel et al. 2016). The following sections discuss how these distinct perspectives are rooted in different constructs of the Cornish landscapes on the sub-societal level.

7.2.1   What is Landscape?

Kühne (2008, 2018) differentiates between different levels of individual landscape construction: the symbolic, aesthetic, cognitive, emotional, and normative. These dimensions provide a good basis for analysing the commonalities and differences between the four narratives identified in Cornwall and their implications for climate adaptation. Table 11 gives a comparative overview of these dimensions of landscape with regards to the four narratives.

On the symbolic level, specific elements of the landscape carry symbolic significance, and the values attached to them differ widely betw een the narratives. Whereas the human-environment interaction narrative focuses on the traces of human dwelling in the landscape and views them as symbolic for the long-standing history of settlement in Cornwall, in the NS narrative specific types of habitats and local species are perceived of symbolic for the sensitivity of the county's ecosystems (cf. sections 5.3.1, 5.3.2). The VB narrative sees the beauty of the Cornish landscapes and their relative unspoiltness as symbolic for Cornwall as a unique county, while in the FL narrative agriculture and renewable energy infrastructure are understood as symbols of efficiency and progress (cf. sections 5.3.2, 5.3.4). Such different interpretations of the same physical features inevitably result in different viewpoints on their relevance for Cornwall's landscapes as well as how they should be dealt with under a changing climate (cf. section 7.2.3).

The narratives also clearly differ in their viewpoints regarding the aesthetic dimension of landscape construction. In the human-environment, natural systems, and visual beauty narratives, the visual attractiveness of certain landscape elements is clearly viewed as worthy of preservation, and thus central to landscape management (see Table 11)). Moreover, the question if and which built structures in the landscape are tolerable is rooted in the aesthetic dimension of landscape construction. Whereas in the visual beauty narrative most built structures are rejected, the HE perspective is accepting of human settlements and traces of human practice if conform with said local distinctiveness (cf. 5.3.1). From the perspective of the FL narrative, the aesthetics of the landscape are only of low importance, and engineered renewables infrastructure or other built features are not seen as problematic (cf. 5.3.4). These different symbolic and aesthetic values attached to certain elements of the Cornish landscapes give these features differing levels of importance in local actors' considerations around landscape management and climate adaptation. Understanding the underlying values of such landscape features gives important insights into why, for instance, certain adaptation measures happily welcomed from one perspective, but are unacceptable from another.

O'Brien & Wolf (2010: 233) underline that successful adaptation to climate change depends on what landscape elements certain actors seek to preserve (cf. section 2.1) (see Table 11). A central question in this context is: what elements and features of the landscape must be protected from negative climate change impacts? In this sense, all implications for landscape management that derive from the perspectives of the narratives fall under this category (cf. also section 7.4). An analysis of a number of particular societal values underlying these symbolic, aesthetic, cognitive, emotional, and normative levels of landscape construction offers deeper insights into the rationales behind the different adaptation approaches; namely specific perspectives on nature, the human-nature relationship, and climate change.

7.2.2 The Human-Nature Relationship and Climate Change

The question which landscape elements are natural and which are human-made is subject to individual and collective ascriptions of meaning (cf. Castree 2013, Kühne 2018) (cf. 3.1). This social constructedness of nature and the human-nature relationship becomes apparent also in the four narratives identified in Cornwall. What is perceived as nature or natural thereby varies between the narratives; likewise so do the perspectives on the relationship between nature and human activity. These viewpoints result in distinct perspectives also on the phenomenon of climate change and how it affects the Cornish landscapes. An analysis of these understandings of nature, human activity, and their interrelations opens up valuable insights into why the local actors lay focus in their adaptation approaches on different elements of the landscape, and which adaptation options they consider or not.

Figure 36 illustrates how the landscape is constructed in the different narratives with regards to these points. From the view of the human-environment interaction narrative, as the title suggests, the landscape is understood as interplay between nature and human activity, one shaping the other reciprocally (see Figure 36, top left) (cf. 5.3.1). Human dwelling stands at the center of this narrative and human-induced change is viewed as integral component of the landscape (cf. Ingold 1993) (cf. section 3.1). This perspective on the landscape is very much one of 'insideness' (cf. Wylie 2007) (cf. 3.4.2). Just as the landscape is understood here as a hybrid between nature and human activity, so is climate change. It is seen as anthropogenically caused and accelerated, and spoken about with a notion of threat. As local communities stand in the centre of this landscape construction, the impacts of climate change are perceived to affect primarily these communities, their practices, livelihoods, and the local economy. Examples for such impacts are damaged buildings near the coast, impediments to road and train infrastructure, problems for fishing communities through damaged slipways, or disadvantages for the tourism sector in form of unstable weather conditions. Interviewees following the HE narrative thereby acknowledge that the rising frequency of events of river and tidal flooding are the result of both physical climate change impacts on the region as well as human (mal)practice in land management, building, and the maintenance of infrastructure (cf. 5.3.1).

In the natural systems narrative, on the contrary, nature and humans are viewed as two separate systems with the natural one under pressure from the human. The natural elements of the landscape are seen as under threat from harmful human activities such as housing developments or pollution from farming (see Figure 36, top right). Moreover, nature is here perceived of as unpredictable and outside the control of humans. Resulting from this are a critique of human presence in the landscape and a feeling of responsibility and stewardship over the natural environment (cf. 5.3.2) (cf. Antrop 2004, Gailing 2014). Interestingly, *what* is viewed as nature and tolerable local wildlife differs from the perspective of the NS narrative. Whereas species that have been local to Cornwall for a long time are accepted as natural parts of the landscape, new 'invasive' ones settling in Cornwall due to changing climatic conditions are met with skepticism and perceived as a threat to 'proper' local wildlife (cf. section 6.2). The perception of climate change in the natural systems narrative strongly differs from that of the HE narrative. The question of the role of human activity is not viewed as central, because changes in the climate are seen as natural and inevitable. Climate change impacts are understood in terms of their effects on the natural elements of the landscape such as local wildlife or runoff regimes. However, climate change is here *not* understood as a threat, but as a natural process that potentially induces positive habitat change (cf. section 5.3.2).

While the perspective on nature in the NS narrative is quite differentiated, interviewees following the visual beauty narrative have a very superficial understanding of nature. Here, the term 'landscape' is used synonymously with 'natural environment', both referring to the "green and pleasant" elements of the local landscapes (I-8: 9) (cf. section 5.3.3) (see Figure 36, bottom left). This view is blind to environmental issues related to the landscape such as farming pollution, the functioning of different landscape elements in tune with each other, or how precisely problems like flooding come into being. The focus here lies merely on the visual attractiveness of a green-looking countryside; a fact that goes hand in hand with low awareness of climate change impacts on Cornwall. As little knowledge exists in the VB narrative of detailed processes in the landscape, the impacts of climate change on Cornwall are largely overlooked from this perspective. Climate change is not perceived of as having negative impacts on the county, and issues such as river or tidal flooding are seen as fixable through small-scale solutions; a perspective reflecting an 'outsider' perspective on the landscape delighted by the beauty of the scenery (cf. section 5.3.3).

Lastly, in the functional landscape narrative nature is conceptualised very differently than in the three previous ones. In the FL narrative the landscape is perceived of as "industrial", a site for food production, and "very, very little [...] 'natural' at all" (I-3: 1) (cf. section 5.3.4). Hence, nature is here understood as *outside* the landscape; the landscape and nature are two distinct phenomena (see Figure 36, bottom right). The natural environment in Cornwall is seen as human-made and domesticated. As the aesthetics of the landscapes do not play a role from this perspective, structures such as renewable energy infrastructure cannot disturb its natural beauty. As thematised by Castree (2013), moreover, nature in the FL narrative is personified and viewed as beyond the control of humans both in a broader sense and at Godrevy specifically (cf. section 5.3.4, 6.2). Based on an understanding of the landscape as primarily human-made, the perception of climate change in the FL narrative centers on its anthropogenic acceleration and its impacts on human activity. Physical climate change impacts from this perspective are viewed in close connection to agricultural production and human settlements. Other than in the HE narrative, however, nature is here perceived of as uncontrollable and climate change as threatening beyond the point where planned landscape management activities can avert its worst impacts (cf. section 5.3.4).

The distinct understandings of nature and climate change highlighted above show that the same physical space can be superimposed with both natural as well as cultural landscape elements, resulting in particular normative stances on their management. In the HE narrative the landscape is actively managed *for* people and communities, and in the VB narrative it is protected for the enjoyment and consumption *by* people. From the perspective of the NS narrative, on the other hand, the separation of the human and the natural system results in a clear imperative for protection of the natural elements of the landscape *from* potentially harmful human interventions. The FL narrative with its focus on effectiveness and production clearly prioritises an efficient, socially sustainable *use* of the landscape (e.g. for renewable energy infrastructure) over the preservation of species or natural beauty (cf. 5.3.4). As outlined above, these differing conceptualisations of nature within the four narratives have implications for the broader adaptation approaches resulting from them. However, they also influence the negotiations of the physical-material relocation activities at Godrevy headland (cf. section 7.4).

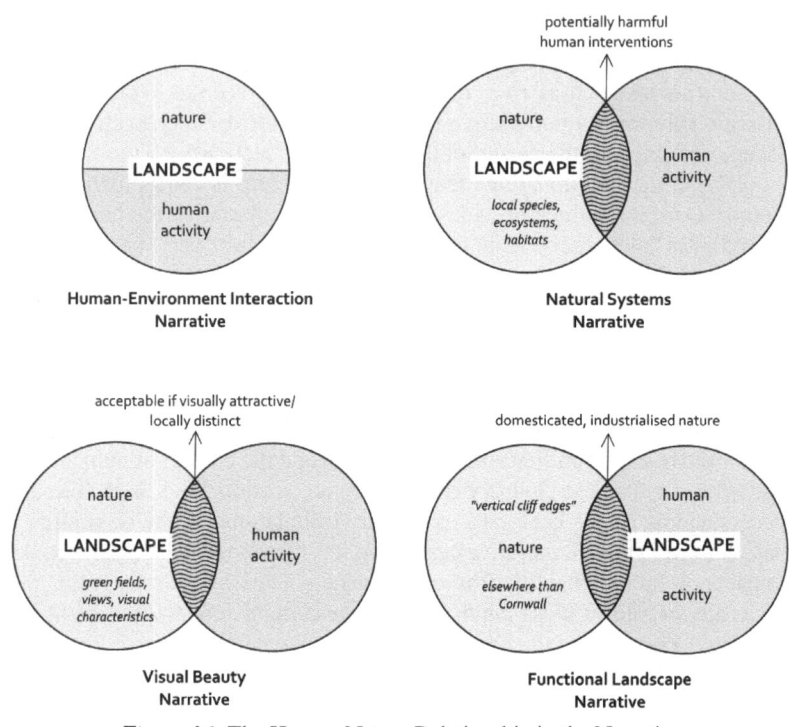

Figure 36: The Human-Nature Relationship in the Narratives
Source: own figure based on interview data

7.2.3   Foci on Communities, Wildlife, Aesthetics, and Efficiency

I showed in the previous sections that the different landscape narratives among local actors in Cornwall have decisive influence on their approaches to climate change adaptation. Differences exist between the narratives regarding the question of what *is* the landscape in general, how nature and human activity stand in relation with each other, as well as how climate change and its impacts are perceived. Whereas the collective place attachments and regional identity of the interviewees lay a shared foundation for adaptation activities, the landscape narratives clearly bear the potential for conflict due to their contrasting foci. Facets of the landscape such as local distinctiveness that feature in all of the narratives thereby constitute superficial common denominators; however, when delving deeper into their symbolic and aesthetic foci, it becomes apparent that the same term by no means necessarily refers to the same phenomenon. From the underlying conceptualisations about the Cornish landscapes and climate change result

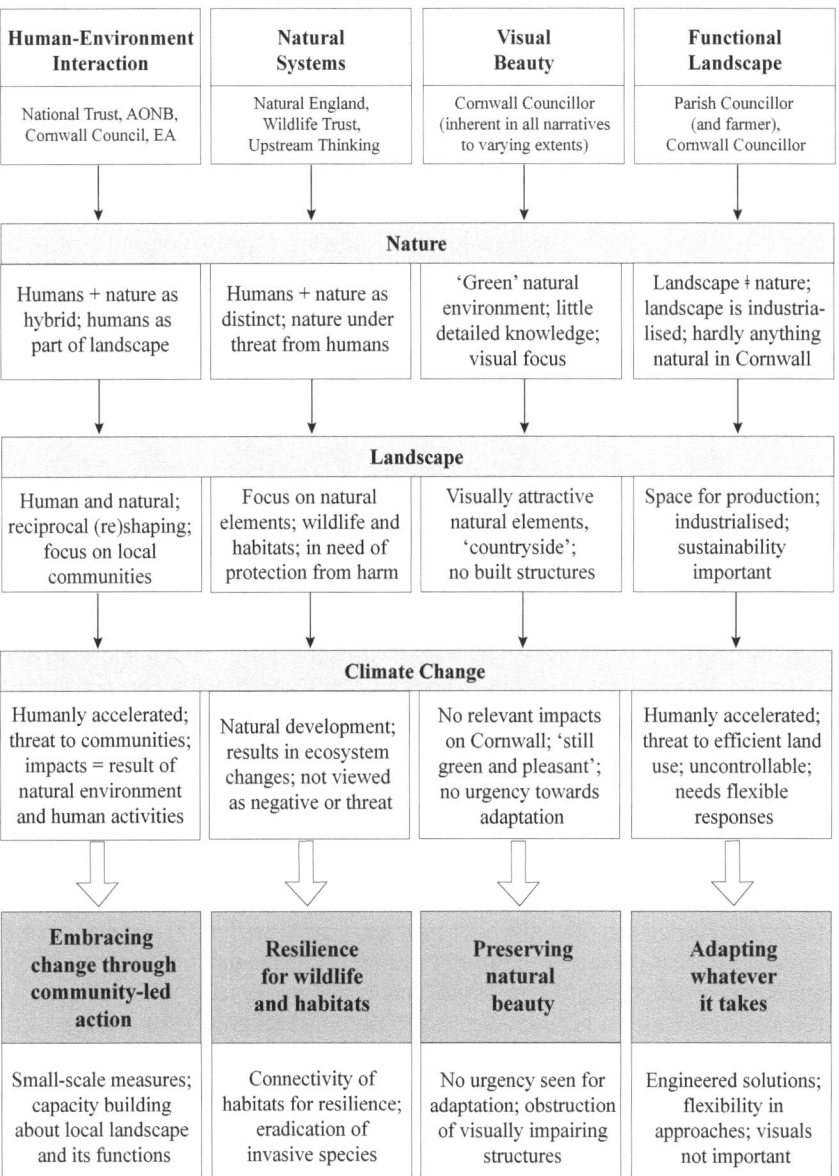

Figure 37: The Narratives' Approaches to Climate Adaptation
Source: own figure based on interview data

very specific foci on *what* must be adapted to climate change and *how* this adaptation should be undertaken. Whereas some actors focus their adaptation efforts on local communities, others center on wildlife, argue for engineered structures, or are very hesitant in general to respond to the changing climate. Figure 37 summarises the points discussed in the above sections, contrasts them between the different narratives, and synthesizes them into the concrete adaptation approaches followed by the local actors in Cornish landscape management (cf. section 5.4). Leaving the sub-societal level of landscape construction and turning to the individual one, the subsequent section is focused on how the interviewees, personally and professionally perceive of the Cornish landscapes, and which tensions arise between their perspectives.

7.3   Between Professional and Personal Perspectives: Three Tensions

Howard (2013: 49) points out that the roles of "professional users of the landscape" are worth investigating, and that the "clash between experts and locals" in this context has so far received insufficient attention (cf. section 2.3.1). As pointed out in chapter 1, one particular focus of this study lies on the roles that local actors in landscape management have in processes of climate adaptation on the ground, and the interplay between their subjective and collective constructions of the local landscapes. The empirical data from Cornwall and Godrevy shows that these individual dimensions of landscape construction are indeed more complex than conceptualized in existing literature so far (cf. 3.4.1). On the one hand, the interview participants can be viewed classically as 'experts' in their respective fields, being vessels of both the narratives and factual information about adaptation processes (cf. section 2.3.2). When entering into deeper analysis of their individual accounts of Cornwall's and Godrevy's landscapes, however, it soon becomes apparent that these actors are situated at the interface of a variety of levels and interests. Leaving the dimension of the sub-societal narratives and taking a closer look at their individual accounts, three particular tensions become obvious: (1) tensions between the scalar, cognitive and emotional levels of their landscape constructions; (2) tensions between their personal and professional roles; and (3) tensions between their organizational backgrounds and the specific local context in which they act.

Tension 1: Individual Landscape Constructions
Taylor (2008: 3) argues that "the ordinary everyday landscape reflects deeply who we are and is a storehouse of private and collective memories". I showed in sections 5.3.5 and 6.2 that, shaped by factors such as childhood memories or leisure time activities, each of the study participants have their very own, intimate

relationships with the Cornish landscapes. When reflecting on the role of expert knowledge and lay perceptions of landscape and place, these findings provide valuable insights for understanding the perspectives of different local actors on processes on climate change adaptation on the ground. As pointed out in section 7.1, what the interviewees' attachments to Cornwall as a place have in common is their intensity and their connection to Cornish regional identity. However, the emotional bonds to the local landscapes articulated by the interviewees differ in scale, level of detail, as well as cognitive knowledge about the landscape. Regarding the cognitive level, Ward Thompson (2013) argues that landscape perception of individual actors is based on "whatever tasks they are currently engaged in and expectations of the future as well as of the past" (ibid: 31) (cf. section 3.4.1). In the case of I-8 (Cornwall Council), for example, a lifestyle of town living, not being concerned with the landscape professionally, and viewing it as "eclectic countryside" and a "rural area" to visit occasionally (I-8: 2; 5) results in a low degree of knowledge also about the functioning of ecological landscape elements or the history of human-made features (cf. section 5.3.3). I-10 (CoaST), on the other hand, speaks in high detail about how the potato plants and apple trees in her garden behave differently in recent years, and how she connects this to the changing climate (cf. section 5.3.2). Similarly, I-10 (Parish Councillor) holds in-depth knowledge about historic flood management features in the landscape near her place of birth and residence in mid-Cornwall (cf. section 5.3.1). These differences in the interviewees' personal landscape perceptions result in decision-making about climate adaptation that is based on very imbalanced levels of knowledge about the landscapes they manage.

Based on their personal relationship with the landscape as well as their cognitive knowledge about it, differences emerge also regarding what Wylie (2007) terms 'insideness' and 'outsideness' regarding the landscape. The purely visual perspective on the landscape by I-8 (Counc'lor) thereby represents an outside perspective connected with a lack of knowledge about ecological processes, a lack of personal attachment, and the absence of activities undertaken in the landscape (cf. section 5.3.3). This unreflected romantisation of the landscape comes with a rejection of built features that would disrupt its visual attractiveness. This is the case, however, at the expense of openness to sustainable landscape management or innovative, engineered adaptation measures. Interviewees with an inside perspective, on the other hand, perceive themselves and their actions as part of the landscape, and comprehend the benefits and negative consequences of human activity in Cornwall (cf. section 5.3.1).

Tension 2: Personal vs. Professional Roles

In addition to the differences between the interviewees' individual accounts of the landscape, a tension shows between their professional and personal roles regarding landscape management. In contrast to the image of the objective, rational and detached 'experts' often described in existing literature (cf. section 2.3.2), most of the local actors participating in this study have a close personal relationship with the landscapes of Cornwall and Godrevy. Some interviewees frequent the site for surfing in their free time, whereas others have grown up in the area or know it from holidays spent there in the past (cf. section 6.2). I-G5 (National Trust) recounts applying for his current job based on the fact that Godrevy was his wife's favourite place in Cornwall, whereas another interviewee speaks about having spent family holidays at the site before becoming professionally involved in managing it (cf. section 6.2). Greider & Garkovich (1994: 4) find in this context that people's individual constructions of the landscape strongly reflect their self-definitions. The empirical data from Cornwall supports this conclusion. Moreover, based on their personal identities and attachments to the landscape, the interviewees often have personal opinions about landscape management that differ from those of the organisation they represent. This becomes obvious at the example of I-13 (NT). Although he speaks about the National Trust's critical stance on renewable structures in their landscapes in his professional capacity, he states that personally, wind turbines are among those elements of the Cornish landscape he would miss would they disappear (I-13: 5/6).

As argued above, the interviewees' knowledge about the landscape is informed by both their professional training as well as their personal backgrounds as hobby gardeners, bee keepers, or bird watchers (cf. 5.3.5). Hence when making decisions about landscape management, they draw on professional knowledge from within their fields of expertise and the organisations they work for, but also on experiential knowledge from years of dwelling in the landscapes they are responsible for. Moreover, the strong emotional bonds to Cornwall and Godrevy as places suggest that for many interviewees, the change and management of landscapes are topics of high personal importance. This assumption is supported by I-18 (AONB) who, when speaking about landscape management in Cornwall, asks: "Why is it complicated? It seems like there's a lot of vested interest in everything that you try to do" (I-18: 20).

Tension 3: Organisational vs. Local Context

I indicated in section 2.3.1 that local actors in landscape and environmental management find themselves at the interface between their organisation's guidelines and policies and the local contexts in which they act (cf. Geoghegan & Leyshon 2014). The findings of this study show that this is the case also when it comes to processes of climate change adaptation. At the Cornish level, this tension exists in the form of a strong regional identity, context-specific challenges with regards to climate adaptation (e.g. the peninsula's topography), and a specific constellation of local and regional organisations involved in landscape management. However, it becomes even more apparent in the case of Godrevy. The National Trust's 'Shifting Shores' strategy, for instance, promotes the implementation of natural solutions and realignment along the coastline, the complicated structures of ownership and land management at Godrevy make this vision hard to implement in practical terms (cf. section 6.3). Moreover, the case of Godrevy shows that despite the legal right of the National Trust to make decisions about their properties self-reliantly and commence a formal planning process, the local actors on the site find it important to reach agreement about the relocation measures *before* entering this formal process. Seeking to maintain positive relationships with the staff of the other organisations involved in managing the site, this search for a compromise takes up large amounts of time and numerous meetings. The personal relationships between the individual local actors, some of them long-standing, add an additional layer of interpersonal relationships to these local discussions. It hence comes down to individual decision-makers on the ground to decide when to follow organisational policy and when to reinterpret and adapt it to the local context; giving the opinions of individual staff at the local level particular weight (cf. sections 6.1, 6.3).

Local actors' personal relationships with the landscape are an important facet of how they construct the places they manage. The findings of this study show that the boundary between 'objective experts' and 'subjective locals' is much more fluid than presented in current literature (cf. section 2.3.2). Moreover, it becomes clear that private relationships with the landscape often stretch into the professional realm, and that personal affection for a place can also be a driver for professional decisions. The implementation of adaptation activities on the ground therefore gets 'filtered' through local actors' personal perceptions of the landscapes they manage. Based on these findings I argue that, when researching people-place relationships, a shift is needed in the current understanding of the role of local 'experts' and decision-makers who act within local contexts. These local actors neither fulfill the classic role of 'locals' holding lay knowledge and making decisions based primarily on their attachments and emotions. Nor, however, can they be viewed as classic 'experts' acting merely on the basis of objective, professional, and rational knowledge. With a hybrid understanding of local actors and their decision-making processes, moreover, comes the need for innovative methodical approaches for researching their roles and actions; a point that I refer to once again in section 7.5.2.

7.4 Adaptation as a Societal Negotiation Process

As argued in section 7.2, the landscape narratives among local landscape man-agement actors in Cornwall give direction for broader approaches to climate ad-aptation. The Godrevy case, on the other hand, shows how these narratives influ-ence the discussions between actors on the ground about the implementation of physical-material adaptation measures. Although small in scale, this case sheds light on what shapes the translation of higher-level adaptation strategies into lo-cal contexts under consideration of the involved actors' perceptions of the local landscapes and Godrevy as a place. These local debates concern the physical-material dimension of landscape construction (cf. 3.4.1), resulting in changes to the external physical space underlying the different actors' landscape construc-tions. In addition, two important issues become apparent from the Godrevy case that shape the implementation of adaptation measures on the ground: on the one hand the particular actor constellation at the local level, their relationships, and the roles of individual staff members within this constellation. On the other hand, the question if a landscape is a common good, who should have access to it, and how. Both these issues have decisive influence on the debates about relocation from coastal erosion on the site, and for the different viewpoints on how the Go-drevy landscape should be altered physically-materially.

The Godrevy thus shows that rather abstract values and perspectives such as the landscape narratives concretely influence negotiation processes about coastal erosion on the ground, and lead to specific viewpoints also on changes to the materiality of the landscape. In the following, I discuss these negotiation pro-cesses with regard to the underlying actor constellation, perspectives on coastal erosion, and viewpoints on the landscape as a commons. At the end of this sec-tion, I synthesise these findings along with those of 7.1 to 7.3 into a place-based framework for local climate adaptation, illustrating the societal factors that shape the implementation of physical-material adaptation measures on the ground.

7.4.1   Coastal Erosion as a Spark for Landscape (Re-)Negotiation

Greider & Garkovich (1994) argue that when the physical-material fabric of a landscape changes or is projected to change, a process sets in among the involved local actors of renegotiation of this landscape. The term 'negotiation' is here not understood in the sense of a formal process of policy negotiation, but in terms of broader processes where dominant and established ways of perceiving and fram-ing the landscape change. Such changes tend to be gradual, occurring slowly over time; however, external triggers such as the rapid erosion of a soft rock coastline can lead to more abrupt shifts in different actors' interpretations of the landscape in a relatively short space of time. This negotiation takes place on the basis of local actors' current constructions of this landscape, the values they attach to certain landscape elements, and the visions they have for its future (cf. section 3.4.3). In the Godrevy case, the environmental change setting in motion such renegotiation is that of accelerating erosion of the cliffs around the headland (cf. chapter 6, introduction). Whereas it is not the aim of this study to analyse the detailed course of these negotiations between different local actors (e.g. in meetings, on the basis of contracts, in consensus-findings processes), a focus on local people-place relationships and the landscape narratives provides valuable insights into the underlying values and perceptions that are the prerequisites of different viewpoints on the relocation activities at Godrevy. In doing so, this study uncovers in depth how the differnent suggestions for physical-material landscape change (cf. section 6.4) are rooted in distinct symbolic constructions of landscape on the basis of the same physical-material space: Godrevy headland.

In line with the findings by Greider & Garkovitch (1994), the Godrevy case shows that the debate about relocating the site's infrastructure in response to coastal erosion sparks a renegotiation of the local landscape, about what it is, and what is should be. This negotiation takes place both within the individual local organisations and between the different local actors. The landscape narratives can thereby be viewed as the prerequisites of such negotiations, but the interaction between the different actors in turn might change their narratives. In this sense, the 'appropriated physical space' (cf. section 3.4.1) is a space for social interaction in two ways: on the one hand because the physical landscape is altered materially on the basis of specific societal values, and on the other because societal interpretative patterns for landscapes are dynamic and might change in the course of the process (cf. Kühne 2018). With regards to adaptation to coastal erosion at Godrevy, this means that the changing climate alters the demands that different actors have to the physical fabric of the landscape; that the ways in which the local actors perceive of these landscapes influence the renegotiation processes; and that adaptation measures change the physical spaces on the basis of which landscapes are mentally constructed.

Kühne (2018) differentiates four key values that guide the negotiation of landscapes: tradition, nature conservation, yield, and emotions (cf. section 3.4.3). At Godrevy, the negotiations about the relocation of infrastructure further inland show facets of all these four points. The local actors' viewpoints are thereby influenced by their underlying landscape narratives, understandings of the human-nature relationship (cf. section 7.2), and practical considerations regarding the local landscape (see following section). The interest of the actors following the HE narrative at Godrevy (e.g. NT, AONB, Visit Cornwall) are driven by tradition-, yield-, and emotionally-oriented values. The conservation of the traditional and natural-looking scenery plays an important role here, but also the facts that the visitors contribute financially to the maintenance of the site, and the high value that is ascribed to people's personal attachments to Godrevy. Organisations like Natural England or the Wildlife Trust, on the other hand, follow the NS narrative and are motivated by nature conservation-oriented values and focus on the protection of local ecosystems and species. Lastly the local farmer, following the functional landscape narrative, is interested in yield-oriented values and the efficient use of his land also under changing climatic conditions. Emotional considerations, on the other hand, do not play a role for him (cf. 6.2).

What makes the Godrevy case particularly interesting is the fact that some actors clearly frame the coastal erosion at the site clearly as climate change-induced, whereas others view it as solely natural process (cf. section 6.2). This difference in perspectives is grounded in the contrasting landscape narratives and conceptualisation of nature, but also in different actors' organisational policies. To the National Trust, the AONB, and the Coast Path Association, the developments at Godrevy are clearly connected to climate change and thus humanly-accelerated. The impacts of the changing climate are perceived in the form of impacts on the human-made features of the site, resulting in the wish to relocate this infrastructure further inland. The focus on climate change as a long-term process thereby automatically lengthens the time frame of adaptation planning, setting focus on projected erosion lines up to 100 years into the future. From the natural systems perspective, on the other hand, the human-made elements on the headland are not viewed as part of the landscape, and coastal erosion is not connected with climate change. As a result, the urgency to intervene into this development through relocation activities is perceived of as very low.

The Local Actor Constellation, Responsibilities, and Scales of Landscape Management
The broader interests that local actors have in the landscape provide insights into the rationales their standpoints in its renegotiation. To comprehend what shapes the ways in which local actors renegotiate landscapes in times of a changing climate, however, it is useful to take a closer look at the specific local constellation of the involved actors, their responsibilities, and, in the Godrevy case, understandings of coastal erosion. Geoghegan & Leyshon (2014) find that in the context of climate change, "uncertain, shifting and unexpected alliances […] form and dissolve between [local] actors in the landscape", and underline "the importance of social networks to these alliances and to the environmental visions that they promote or resist" (ibid: 644). At Godrevy, this actor constellation is shaped by the different landscape designations that overlap in the area, the users of the local landscape in form of businesses and visitors, as well as the ownership structures on the headland. When entering into dialogue with the local actors at

Godrevy, a very nuanced picture of their constructions of the local landscape comes to light that reveals quite contrasting perceptions by the different organizations involved in its management. The interviewees representing the National Trust, the AONB, Visit Cornwall and the Coast Path Association, as well as the owner of the local surf school follow the human-environment interaction narrative. In focus of their landscape constructions stands the headland as a visitor destination, the activities people undertake on-site, and their emotional attachments with Godrevy as a place; resulting in the imperative to maintain access to the landscape also under changing environmental conditions (cf. 7.4.2). Deriving from rationale of preserving the landscape for future generations, especially the Trust's approach to both landscape management and climate adaptation is thereby a long-term and forward-looking one with a time horizon of up to 100 years (cf. section 6.1). The natural systems narrative, on the other hand, is represented by Natural England, the Wildlife Trust, Cornwall Seal Group, and the Towans Partnership. From this perspective, the clear imperative for landscape management at Godrevy is the protection of local species, the dune habitats, ecosystems, and coastal heath on the headland. Nature is perceived of as fragile and in need of protection, but much rather from the human uses of the landscape than from the changing climate and its impacts. The urgency for adaptation to coastal erosion is therefore seen as low, because the habitat changes induced by the eroding coastline are viewed as natural and positive. Natural England thereby has a temporal planning horizon similar to that of the National Trust, applying a long-term vision to the management of the site. The Towans Partnership, however, is very reactive in their landscape management and focus on small-scale incidents such as damage to beach access points or dog litter (cf. sections 6.1, 6.2).

With regards to the power relations involved in the renegotiation of the Godrevy landscape, there are clear differences between the local actors. The Trust as the main land owner thereby plays a predominant role in leading the local discussions. Having statutory veto rights in the formal planning process, on the other hand, Natural England and the AONB act from a position of power towards the Trust and could obstruct their plans at a later stage. Seeking consensus with these organisations therefore is crucial for the success of the Trust's relocation plans. The fact that the Trust engages in consensus-finding also with the other local actors is owed partly to the organisation's policy of participation, and partly to the personal initiative of the site's General Manager. Importantly, it is in line with the Trust's perspective on landscape as a space of human interaction, human practice, and personal attachments (cf. section 6.2). This focus on finding a solution for relocation that suits all involved actors in the area, although lowering the likelihood of obstruction during the planning process, lengthens the time frame of the local discussions about adaptation to coastal erosion significantly.

The Proposed Solutions for Physical-Material Landscape Change
From a superficial perspective, there seem to be numerous solutions for the relocation of infrastructure at Godrevy. When entering into in-depth dialogue with
the involved local actors, however, distinct viewpoints emerge on the landscape
and the priorities of its management that significantly limit the options the National Trust has for these relocation activities. As highlighted in section 6.4, the
interviewees at Godrevy propose five distinct options for physical-material alterations that are rooted in their respective landscape narratives and foci of landscape preservation in the area. Different elements of the external physical space
are thereby assigned different symbolic meanings, making changes to them more
or less negotiable for the different actors. The National Trust, for instance, assigns the agricultural land at Godrevy little symbolic value, and proposes to relocate the car park onto current fields and the road through a protected SSSI site.
However, these fields are of high value to the local farmer, and the SSSI site is
cared for by Natural England who strongly hesitate to re-draw its outline. Interestingly, the interviewees by Natural England propose to relocate the car park to
St Gothian Sands; a Local Nature Reserve looked after by the Cornwall Wildlife
Trust, but outside the responsibility of Natural England. What becomes clear
from this is that the spatialities of landscape designations and ownership structures result in the preference of protecting certain areas of the landscape, but
'sacrificing' another ones. The different viewpoints on access to the headland
(see below), the separation of the landscape in "ours" and "theirs", along with an
inherent NIMBYism[19] regarding the relocation of the infrastructure, shows
clearly the importance of comprehending in depth the underlying people-place
relationships and lines of argument when trying to understand the local actors'
suggestions for physical-material changes to the Godrevy landscape (cf. section
6.4). Whereas most of the local actors state a preference for working with natural
processes and are sympathetic towards the Trust's relocation plans, it shows that
in practical terms the local structures of land ownership and landscape designations, the different constructions of the Godrevy landscape, and resulting contrasting perspectives on access to the headland significantly complicate this endeavor.

7.4.2   The Critical Question: Access to the Landscape?

O'Brien (2009) argues that understanding the societal values that underlie adaptation decisions is particularly important when the activities proposed by one
group impact negatively on what another group values (cf. section 2.2). The actor
constellation at Godrevy and their ways of managing the landscape alongside

---

19   NIMBY = 'not in my back yard', cf. DEVINE-WRIGHT (2009)

each other are stirred up by the pressing need to adapt to coastal erosion. The fact that this erosion is connected with climate change by only some of these actors thereby constitutes a point of conflict, and organisations that have been working harmoniously alongside each other for decades begin to challenge each other's perspectives on the local landscape. From these locally contextualised contrasts in landscape construction in combination with the different spatial and temporal management scales suddenly arises disagreement about how to respond to the changing climate. For whom is the landscape managed and, ideally, how? What should be preserved? Is adaptation to coastal erosion necessary, and in which form? These viewpoints crystallise at Godrevy in a particular local debate that is focused on the question if Godrevy is a common good or not, and who should be allowed to access the headland. Rooted in the normative level of landscape construction, this topic has clear implications also for landscape management in practical terms.

The National Trust clearly understands Godrevy headland as a common good. The infrastructure that is being relocated can thereby be viewed as a material common good, whereas the enjoyment of the site, its aesthetic qualities as well as people's attachments to and memories of Godrevy are immaterial ones (cf. section 3.3). The staff of the Trust emphasize the significance of Godrevy as a visitor destination for both locals and tourists from further afar, thus clearly managing the site *for* people. Of particular importance in this context are the personal relationships these visitors have with Godrevy headland, either making use of the landscape on a daily basis or being attached to the place on the basis of repeated holiday stays. The organization's slogan "Forever, for everyone" thereby underlines its primary goal of landscape management: to enable access to the site for people with both good and bad mobility (National Trust 2017c) (cf. section 6.1). In this context, especially the views of the sea and across St Ives Bay are perceived of by the interviewees of the National Trust as integral to the Godrevy experience. Resulting from this is a strong emphasis of the Trust on making it possible also for elderly and less mobile people to visit the site. The large car park on top of the headland therefore is a crucial component of the local access strategy. The symbolic value given to the visuality of the landscape and its enjoyment by all visitors results in the non-negotiability of maintaining physical access to the headland. Maintaining access to Godrevy and the local beaches is not only important to the National Trust as the main land manager, however, but also to other local organizations and businesses. The owner of the surf school, for instance, relies decisively on visitors having access to the site and the possibility to park their cars. Although agreeing with the National Trust about Godrevy being a common good, the AONB Partnership, on the other hand, strongly criticize the location and negative aesthetic impact of the current car park on the top of the headland. Both the natural environment and the landscape are thus managed from an anthropocentric perspective for people, either for the sake of their emotional attachments or for visual enjoyment and recreation (cf. 6.3.1).

The importance given by the National Trust and other local actors to enabling access to Godrevy headland is opposed by the perspective on the local landscape from the natural systems narrative. From the perspectives of Natural England, the Cornwall Wildlife Trust as well as the Towans Partnership, Godrevy is perceived as semi-natural coastal heathland that shelters rare species in need of protection from negative human interventions. Mutton Cove in the north of the headland, moreover, is a major haul out sites for Grey Atlantic Seals in south west England. Therefore, the interviewees from Natural England and the Towans Partnership wish to limit the amount of people coming to the site. Despite the importance of the site for both locals and visitors, landscape management from this perspective must put wildlife and habitat protection first. After all, so the argument, from the high visitor numbers arise negative impacts on the landscape such as the disturbance of sensitive wildlife. Regarding the relocation of the car park and road away from the cliff edge, the question resulting from a natural systems view on the Godrevy landscape is whether it is necessary *at all* to provide access for visitors and cars to the coastline. More limited space for car parking in this context would constitute a welcomed control mechanism for how many people can visit the site at a time (cf. section 6.3.1). The question of access at Godrevy therewith opens up a much wider discussion between the involved actors about whether to generally tolerate people in the landscape or not, for reasons of both wildlife protection as well as preservation of the site's natural beauty. Investigating the actors' distinct viewpoints thereby makes it possible to comprehend which elements of the Godrevy landscape they seek to protect, and what features are non-negotiable for them when it comes to the relocation of physical structures.

Practical Considerations and Ownership Structures
The incentive of the National Trust to relocate existing infrastructure away from the cliff edge is grounded in their coastal management policy 'Shifting Shores', which lists "relocate[ing] people and buildings away from high-risk areas" as one of the key strategies (National Trust 2008: 6). Aside from the debate about whether there should be access to the landscape for people and cars or not, this relocation is not easily done on the ground in practical terms. In addition to the normative question of access to the landscape, the structures of land ownership and management at Godrevy complicate the implementation of physical landscape change (cf. 6.3.2). Geoghegen & Leyshon (2014: 639) find similarly in their Lizard Peninsula case study that when it comes to climate adaptation, local

actors are faced with "an area that is geographically diverse with coasts, valleys and heathland, as well as split between local residents, second-home owners and visitors with differing aspirations". Although both the National Trust's 'Shifting Shores' strategy as well as the organization's local staff at Godrevy would prefer to relocate the access road and car park to locations directly inland from the current ones, in practical terms the choice of alternatives is significantly limited by the local structures of land ownership and management. A braid of National Trust property, tenanted farmland, privately owned land, as well as protected wildlife sites significantly reduce the options the National Trust has for moving the infrastructure further inland to maintain access to the headland.

7.4.3   A Place-Based Framework for Local Climate Adaptation

Geoghegan & Leyshon (2014) find that on the Cornish Lizard peninsula, adaptation activities on the local level are made up of "the complex relations between institutional priorities, people and place [which] are negotiated" in their local context (ibid: 644). Throughout this study, I showed with a high degree of detail that the process of climate adaptation must indeed be understood as a negotiation processes between local actors against their collective, sub-societal, and individual constructions of the physical spaces affected by climate change. In this context, tensions arise between higher-level strategies and the requirements for adaptation at the local level; broader organizational guidelines and local contexts; the viewpoints of different actors on the ground regarding their landscape constructions and perceptions of nature and climate change; as well as the personal and professional roles of individual staff members responsible for decision-making. Especially the Godrevy case shows that varying conceptualizations of the same place lead to different approaches to climate adaptation. The relocation of infrastructure further inland is thus clearly a societal negotiation process, complicated by different perspectives on erosion and the question of access to the landscape. The four landscape narratives result in distinct standpoints on what elements need protection and what changes are non-negotiable; additionally, the consensus-finding process on the ground is complicated by both normative and practical considerations.

On the basis of this study's focus on people-place relationships and the findings from Cornwall and Godrevy, I find that adaptation processes translate into local contexts through five what I term 'filters' on the collective and individual level. A structured analysis of these filters in studies investigating climate adaptation processes decisively aids in understanding the societal values that shape local actors' decision-making and in comprehending how specific adaptation measures come into being. Such assessment of the values that 'filter' larger-scale adaptation strategies into local contexts is not only useful in the Cornwall example, but in any other case study that investigates the local embeddedness of processes of climate adaptation and landscape management. I therefore synthesise the findings of this study in a Place-Based Framework for Local Climate Adaptation, shown in Figure 38[20], which constitutes a heuristic device for use in structured assessments of the factors that shape local climate adaptation processes. Each of the proposed filters should thereby not be viewed as structural elements only, but also as elements informing individual and collective perceptions and constructions of the landscape. The legal regulations framing adaptation processes in their local and national context, as well as underlying government acts and the formalities of planning processes, were not in focus of this study and are excluded from detailed description in the framework (see dotted line).

Based on the findings of this study, the place-based factors and societal values that shape such processes locally are therefore:

1. **The specific actor constellation on the ground:** their duties and privileges, their relationships with each other; as well as the spatiality of their responsibilities (e.g. in form of landscape designations);
2. **the individual perspectives of local professional actors** *(individual level of landscape construction)*: their individual landscape constructions; personal relationships with and attachments to the local landscape;
3. **the specifics of regional identity** *(societal level of landscape construction)*: the distinct role of place and landscape; regional self-perception; notions of conservatism and innovation; particularities of geographical locations;
4. **narratives about the local landscapes** *(sub-societal level of landscape construction)*: distinct perspectives by local actors on what the landscape is; on the human-nature relationship; on climate change and its impacts; as well as on suitable adaptation measures;
5. **structures of ownership and landscape management** on the ground that restrict the implementation of physical-material adaptation measures.

It was the research objective of this study to investigate how people-place relationships influence local processes of climate adaptation, and how these processes then result in the implementation of physical adaptation measures. With the above discussion of the societal, sub-societal, and individual levels of landscape construction and place attachments, I showed that context-specific

---

20   As shows from the empirical findings, beyond the vertical relationships between the national and local level of adaptation activities, there are numerous horizontal interlinkages between the different aspects of the five filters. These horizontal links are not shown here for the sake of ease of interpretation.

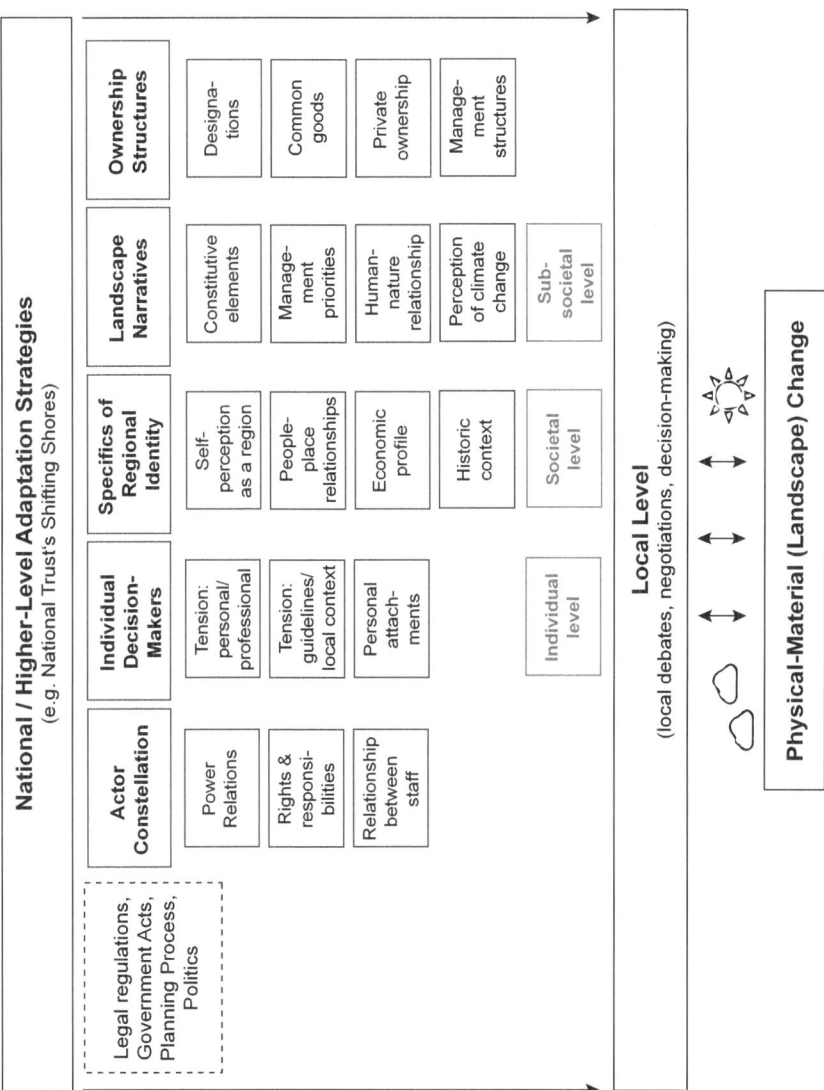

Figure 38: The Place-Based Framework for Local Climate Adaptation
Source: own graphic based on research findings

people-place relationships in fact have an important influence on how adaptation to climate change and – in this case – coastal erosion is approached at the local level. Higher-level adaptation strategies, for instance by governments but also by nation-wide organisations like the National Trust, thereby get mediated on the ground by a number of what I term 'filters' that translate their often unspecific targets into specific local contexts. In the sense of a values-based approach this study thus shows that climate adaptation is indeed shaped by societal values (cf. section 2.1), and that the societal constructedness of the places affected by the changing climate is a crucial facet of these values. The use of the Place-Based Framework of Local Climate Adaptation, I argue, is a valuable approach for investigating what shapes societal processes of adaptation to climate change on the ground.

7.5 Critical Reflection of the Research Process

This study was undertaken with the aim of investigating the role of people-place relationships in processes of climate adaptation at the local level. To approach this research objective, I chose an inductive, interpretative qualitative approach to data collection and analysis, using the constructivist landscape concept as the main theoretical framework. In this section, I reflect critically on this choice of theory, methods, and the overall research design, hinting also to the limitations of this study. Moreover, I assess my own role as a researcher in this process.

Theoretical Framework
Revill (2016: 4) argues that in research, "theorizations, concepts and metaphors have to be cast upon the world as points of departure rather than as attempts to control, define, and capture" the societal processes under investigation. The theorization drawn upon in this study is that of social-constructivist landscape research. Both the concept of landscape must thereby be understood as exactly this: concepts or, in other words, auxiliary constructions aiding the researcher in comprehending the complex societal processes that stand in focus of her inquiry. In this sense, this scientific perspective constitutes a societal construction of landscape just as those by the local actors participating in this study do; the former understood here as a vehicle for uncovering and interpreting these actors' perceptions of the places affected by climate change, their conceptualisations of nature, and their perspectives on climate change. As Gieseking et al. (2014) conclude, "landscape remains a useful term and tool for thinking about the interface between humans ('culture') and the non-human world ('nature') at a range of scales". The concept by Kühne (2018) in particular thereby enabled me to grasp the collectively shared, sub-societal, and individual dimensions of local actors'

landscape perceptions, as well as the commonalities and difference between them. It thus constituted a valuable tool in analysing the interview data.

When reflecting on the use of a concept as broad as that of landscape, however, I must allude to the conceptual baggage it comes with emerging from its long and vast disciplinary history in human geography, as well as the multitude of topics and phenomena for that it is an umbrella term. When trying to familiarise myself with the many understandings of landscape in German-language human geography, let alone the British debate, I soon became aware that grasping the entire range of connotations of the concept in this discipline clearly exceeded the scope of this study. I nonetheless argue that the concept of landscape, in the way I used here, constitutes a very useful tool for shedding light on people-place relationships in the context of climate adaptation processes.

Research Design and Methods
Especially the methodical use of walking interviews and the analytical use of a narrative approach resulted in rich data and valuable insights into local actors' people-place relationships. The walking interviews thereby brought useful results regarding the data I could obtain about different landscape constructions and adaptation approaches at Godrevy, but also with regards to the specific roles of the interviewees as situated between the personal and professional realm. Considering the landscape narratives and perceptions of climate change, the active 'being-in-the-world' (cf. Wylie 2009: 278) with the participants provided valuable physical-material and sensual prompts for their replies to the interview questions, enriching the data about their perspectives significantly. On the downside, however, it is exactly these weather conditions and the potential crowdedness of the site under investigation that might render the interview process difficult, requiring a certain amount of experience and skill on the side of the researcher. Moreover, the rich data that I could collect at Godrevy shows that, other than suggested in existing literature (cf. 4.2.1), walking interviews are a valuable method not *only* to inquire about people's personal stories of and attachments, but also about their professional roles and knowledges. Particularly useful in this context was also the combination with a narrative approach to data analysis, which aided in the identification not only of individual perspectives, but also storylines and overarching narratives regarding the societal and sub-societal levels of landscape construction (cf. 3.4.1).

Role of the Researcher
Chase (2005: 657) outlines that "as narrators, […] researchers develop meaning out of, and some sense of order in, the material they studied; […] they 'narrate' results on ways that are both enabled and constrained by the social resources and

circumstances embedded in their disciplines, cultures, and historical moments". As Chase emphasizes here, bias is an inherent component of qualitative research and cannot be avoided (cf. also Lewis 2003: 13). Therefore, as a researcher one carries specific responsibility to remain self-aware and reflective during the entire process of data collection, analysis, and interpretation.

During the process of data analysis, what I had to set aside was my *very own* construction of the Cornish landscape. The three-step approach to narrative analysis by Feldman et al. (2004) (see above) was a great help in this: by analysing the data very closely to the text, and thus closely to the interviewees' accounts of the Cornish landscape, it was possible to set aside my own ideas about this landscape and 'listen' to my interview data with reduced bias. My personal relationship with the Cornish landscapes thereby influenced how I related to local actors and how they related to me: On the one hand, a shared appreciation of the landscapes' attractiveness provided common ground for dialogue during the interviews. On the other, my positionality towards the landscape always clearly remained that of a visitor looking in from the outside. This obvious outsider perspective resulted in very detailed accounts of how the interviewees perceived of the landscape, attempting to explain to me even those facets of Cornwall's landscapes that are taken for granted by those who interact with them on a daily basis.

On the basis of the above discussion of the empirical findings of this study and the research process, in the next and final section I draw conclusions from the previous seven chapters about the role of people-place relationships for climate adaptation at the local level.

8 Conclusions

At the beginning of this book stands a quote by Bugler & Palin (2017: 1) who conclude that adaptation to climate change is, "to a large extent, a local process" and therefore needs to be "sensitive to local contexts". After having conducted empirical research in Cornwall and at Godrevy, I find that a focus on people's perceptions of the physical spaces where climate change takes place provides an optimal focus for research on climate adaptation. The overarching research question of this study was therefore:

How are processes of climate change adaptation influenced by the social constructedness of the affected places, and how in turn does an abstract phenomenon like climate change become evident physically-materially in a socially constructed place through concrete societal actions?

By drawing on constructivist landscape theory for addressing this question, I could show that the relationships that local actors have with the places and landscapes they manage (and in which they might also live) indeed have a decisive influence on how climate adaptation is approached generally, and on how local debates about adaptation to an eroding coastline more specifically. In the following sections, I summarise the key findings of this study; outline the contributions it makes on a theoretical and methodical level as well as its implications for practice; and point out avenues for future investigations in the field of place-based and locally-contextualised climate adaptation research.

8.1 Key Findings

Adaptation to climate change is inherently local (cf. Agyeman et al. 2009, Agrawal 2010, Adger et al. 2011, Bugler & Palin 2017). Whereas strategies for adaptation can be drafted and issued on a variety of levels from the international to the local, both the impacts of the changing climate and the implementation of physical-material measures inevitably take place 'on the ground' and in local contexts. However, it is precisely the consideration of this ground which is missing in existing literature about climate adaptation (cf. Geoghegan & Leyshon 2012, Köpsel et al. 2016). Although growing attention is paid in the field of human geography to the societal values that shape adaptation processes, the perceptions and interpretations that different actors have of the physical-material spaces affected by climate change impacts are mostly neglected in empirical

© Springer Fachmedien Wiesbaden GmbH, part of Springer Nature 2019
V. Köpsel, *New Spaces for Climate Change*, RaumFragen: Stadt –
Region – Landschaft, https://doi.org/10.1007/978-3-658-23313-6_8

investigations (cf. Devine-Wright 2015, Tomaney 2015). I argued throughout this study that the assessment of place attachments among local groups and individuals is a good starting point for developing a place-based perspective on adaptation; however, as I could show at the example of Godrevy, shared affection for a place alone does not result in agreement on how to adapt it to the impacts of climate change. Comprehending what shapes societal adaptation processes requires closer examination of the ways in which actors at the local level perceive of the places that they manage. Here conceptualised in terms of landscape constructions, this study shows that differing societal constructions of place importantly shape approaches to climate adaptation, and local debates about responding to coastal erosion (cf. chapters 5, 6).

On the societal level, climate adaptation in Cornwall is influenced by collectively strong place attachments, and a distinct regional identity closely linked to the county's landscapes and torn between conservatism and innovation. Subsocietally, different actors in Cornwall construct the region's landscapes in terms of (1) human-environment interaction, (2) natural systems, (3) visual beauty, and (4) functional spaces for production (cf. section 5.3). Hand in hand with these narratives go specific values attached to certain landscape elements, particular understandings of nature and the human-nature relationship, as well as differing approaches to climate adaptation deriving from them (cf. section 7.2). The HE narrative results in a focus of adaptation on community-lead, small-scale activities and the re-attachment of the local population with the landscapes in which they live and work. From the NS perspective, climate adaptation addresses habitat connectivity and environmental resilience by protecting local species and supporting adaptation measures that work with nature, not against it. Due to its primarily aesthetic focus, the VB narrative implies a superficial view on the Cornish landscapes that results in a lack of urgency towards adaptation; any physical implemented, however, should be sensitive to traditional building styles and the region's natural beauty. Lastly, from the FL narrative derives an approach to climate adaptation that sets aside visual considerations in favour of (engineered) solutions with high efficiency to protect local communities and mitigate CO_2 emissions (cf. 5.4.2).

The investigation of the different narratives and resulting adaptation approaches on the Cornish scale provides valuable empirical evidence of the role of landscape constructions and people-place relationships for distinct approaches of climate adaptation. However, it stays fairly abstract. The case study of Godrevy, on the other hand, illustrates how such contrasting landscape constructions influence the implementation of concrete, physical-material adaptation measures on the ground. Faced with accelerating coastal erosion and a visitor infrastructure under threat, the local actors find themselves in a time-consuming negotiation process about the ideal solution for its relocation. Different interpretations of the same physical space, namely Godrevy headland, thereby connect the interviewees' accounts of the landscape closely to its material fabric and result in opposing viewpoints on the relocation measures (cf. 6.2, 6.3, 6.4). Although not yet implemented, these measures will in turn alter the physical-material space that underlies the different landscape constructions, making these constructions both embedded in their local contexts as well as fluid and dynamic.

What the Godrevy case shows clearly is the fact that, without an in-depth inquiry into the local actors' distinct perceptions of the area's landscape and their rationalities for relocation, the headland superficially appears to be a physical space offering a variety of options for building a new road and car park. Once entering into dialogue with these actors, however, it soon becomes clear that these options are decisively limited by the societal constructions of Godrevy as a place. These differing interpretations of the local landscape crystallise in the debate about Godrevy as a commons and the question of who should, and should not, have access to it. Underlying this debate are specific understandings of the relationship between nature and human activities, but also of climate change and coastal erosion (cf. section 6.2, 6.3, 7.4). Another layer of complication is added by ownership and land management structures which, in themselves, are results of societal processes of the construction of physical spaces and their superimposition with, for instance, landscape designations (cf. 6.3.2).

On the individual level, this study lays open the specific roles that local actors in landscape management occupy on the ground. Situated at the interface between larger-scale policies and specific local contexts, these actors find themselves under three tensions: a tension between their own landscape constructions and those of the colleagues they engage in decision-making with; a tension between their personal attachments to Cornwall's landscapes and their professional roles as landscape managers; as well as a tension between their organisational backgrounds with respective management strategies and the local contexts in which they act (cf. sections 5.3.5, 6.2, 7.3). These empirical findings suggest a re-thinking of the conceptualization of 'experts' and 'locals' in existing literature (cf. 2.3); after all, as this study shows, the individuals making decisions about landscape management and the implementation of adaptation measures on the local level are neither can be viewed as 'locals' associated with decision-making based on attachments and emotions, nor are they entirely objective 'experts' (cf. 7.3).

I synthesise the findings of this study in a Place-Based Framework for Local Climate Adaptation (see Figure 38). On the basis of the empirical data, this framework proposes five 'filters' through which larger-scale adaptation strategies are translated into local contexts: the actor constellation as well as the individual decision-makers on the ground; the specifics of regional identity in a given place; existing narratives about the local landscapes; and, on a practical level, distinct structures of ownership and land management (cf. section 7.4.3). As a heuristic device and combined with a qualitative methodology, this framework is an auxiliary tool for uncovering the unspoken, taken-for-granted perceptions and lines of argument that underlie different perspectives on local places and adaptation measures, setting particular focus on individual actors' viewpoints on and interpretations of these activities. The application of this framework opens up a perspective on climate adaptation as a societal negotiation process in a field of tension between larger strategy and policy frameworks, local organizations with their vested interests, individual decision-makers, and local populations. As local climate adaptation processes are decisively shaped by societal values (cf. Agyeman et al. 2009, Bugler & Palin 2017), laying open the inherent place-based values guiding local decision-making is a useful approach to comprehending the renegotiation of landscapes in times of a changing climate, as well as the factors that hinder and support effective consensus-finding in this context.

8.2 Contributions of this Study

Beyond the empirical findings presented in the previous section, this study makes contributions on the empirical, theoretical, and methodological level, and has practical implications for processes of landscape management and climate adaptation at the local level; contributions that I outline throughout this section.

Empirical Contributions
By presenting empirical data from Cornwall, I provide detailed findings on landscape constructions on the societal, sub-societal, and individual level. These findings do not only widen the evidence base for understanding how landscapes are constructed on different levels of society generally, but also highlight the implications of these constructions for a specific issue – in this case multi-actor processes of adaptation to climate change. By setting the focus on people-place relationships and their implications for physical-material landscape change, I address an important research gap in the context of values-based climate adaptation research and demonstrate the suitability of a landscape approach for doing so. Especially the Godrevy case study provides a tangible example of the societal renegotiation of landscapes prompted by accelerating coastal erosion. This case study shows that different mental constructions of local landscapes indeed result in distinct perspectives on their physical-material reshaping. In addition, the empirical data provides impetus for re-thinking the conceptual usefulness of the

long-standing divide in human geography between natural and cultural land-scapes by showing that on the ground, the same physical spaces are indeed over-laid with mental constructions as both natural and physical landscapes, or hybrids of the two.

Especially the focus on individual actors in landscape management delivers in-teresting contributions to the understanding in existing literature of the roles of 'experts' and 'locals'. Transcending the dichotomy between these categories, these individuals act at the interface between their professional and personal roles, as well as between larger-scale organisations and local contexts. As I could show based on this study's empirical data, their decisions are, to differing ex-tents, influenced by both professional as well as experiential knowledge. To un-derstand the knowledge and values underlying decision-making processes on the ground, I argue, a re-thinking of the current conceptualisation of these 'experts' is necessary to fully grasp their role in the implementation of climate adaptation measures, but also in the wider context of local landscape and environmental management processes.

Theoretical Contributions
On the theoretical level, this study is innovative in applying a landscape perspec-tive to an investigation of local climate change adaptation processes. By concep-tualising people-place relationships in the form of individual, sub-societal and societal landscape constructions, it provides a structured analysis of local actors' understandings of the landscapes that are affected by climate change impacts, the changes resulting from these impacts, and the proposed responses to these changes based on distinct, co-existing perceptions of the same physical spaces. A social-constructivist approach, as I could show, thereby results in valuable in-sights into people's reasoning behind their responses to climate change. Moreo-ver, I enrich existing literature by linking the debate about climate adaptation in its local context with the topic of landscape as a common good; a topic that is likely to become more important in a future with accelerated coastal erosion, but also changes to other publicly-used physical spaces affected by climate change impacts such as flood plains, woodlands, or public infrastructure (cf. e.g. Dawson 2015). Beyond the issue of mere physical accessibility to these spaces, different viewpoints to the normative question of who *should* have access to these spaces and how importantly shapes local debates about responding to the impacts of climate change.

With regards to concrete processes of climate adaptation at the local level, this study highlights the importance of considering societal constructions of place and landscape for understanding what shapes local decision-making in this con-text. It becomes clear from the case studies of Cornwall and Godrevy that especially in locations where place attachment is strong and the local landscapes play a central role in regional identity, contrasting constructions of these landscapes importantly influence how climate adaptation is approached. Moreover, these emotional bonds to the landscape result in high vested interests on the sides of all actors involved in decision-making, potentially leading to conflict about the implementation of physical-material measures. As I could show at Godrevy, local climate adaptation must consequently be understood a process of negotiation between a variety of local actors, guided by equally many societal values, that leads to the reshaping of physical spaces based on these actors individual interpretations of these spaces. In the context of values-based climate adaptation research, this study makes a valuable contribution by emphasising those societal values that are place-based and connect local actors to the materiality of the spaces they manage through a vast set of interpretations and ascriptions of meanings; meanings that often contrast and result in very different conclusions on how to address the impacts of climate change. While considerations of technical feasibility or cost-efficiency certainly are important factors in local adaptation processes, to understand such processes on the ground we must comprehend how actors perceive of the places where they live and work, what elements of their landscapes they seek to protect, and what measures are unacceptable from their contextualised perspectives. By developing a Place-Based Framework for Local Climate Adaptation, I propose a heuristic tool for the assessment of such people-place relationships and place-based values; thus opening up an avenue for grounding adaptation processes in their local contexts and systematically analysing the roles of societal, sub-societal, and individual constructions of place and landscape in decision-making about the (re)shaping of physical spaces in response to environmental change.

Methodological Contributions
On the methodical level, this study's research design as a small-scale case study embedded in a larger regional context highlights the added value of in-depth qualitative research into societal processes of landscape and environmental management. A focus on the individual accounts of those local actors responsible for making adaptation decisions on the ground is thereby rare in existing literature. Laying open their individual and collective relationships with the places they manage provides valuable insights into what shapes their decision-making, and how the implementation of physical-material measures comes into being in the interplay of local organisational staff, political representatives, farmers, and members from the wider public such as local business owners.

Importantly, moreover, this study highlights the added value of walking interviews for researching people-place relationships and perceptions of the local landscape on the one hand, and the suitability of a narrative approach for analysing such different viewpoints on the other. Whereas often used to address the issue of emotional attachment in existing studies, I could show that the application of a non-sedentary approach to interviewing provides rich data and evokes an in-depth reflection of the local landscapes and its management by the interviewees also with regards to more factual, professional knowledge. The material elements of the physical landscape thereby serve as important 'prompts' for addressing topics like landscape change and climate change impacts, but also personal relations with the landscape. Collecting data on both the participants' personal *and* professional roles, this joint 'dwelling' in the landscape is therefore a useful tool for researching societal processes connected to specific physical spaces, and for uncovering these places' societal constructed-ness. Moreover, I could demonstrate the usefulness of the analytical tool of narratives for structuring this rich body of data and analysing the accounts of local landscapes on the societal, sub-societal, and individual level.

Implications for Practice

Greider & Garkovich (1994: 14) argue that the use of a constructivist landscape approach for analysing societal negotiations of landscape and environmental change can improve the effectiveness of local management processes by showing "that the environment has multiple meanings, […] and that changes in the environment can […] require a renegotiation of the meaning of both themselves as people and their relationships to the environment". In this sense, laying open the unspoken meanings conferred to different elements of physical creates the opportunity to challenge and discuss local actors' implicit assumptions about the places they manage, and to find common ground between their perspectives that can serve as a basis for consensus-finding. In the context of Cornwall, such common ground can be found in the collectively strong attachments to the region as well as the appreciation of its natural beauty. These attachments to local distinctiveness and the pride of the Cornish population of their region could serve as a mobiliser for adaptation action both on the side of local landscape managers as well as local communities. Where the landscape constructions uncovered through empirical research delve deeper than the popular stories about mining, surfing, and natural beauty, however, open dialogue about the differences in place constructions can foster understanding between local actors of contrasting perspectives on proposed adaptation measures, and potentially resolve tensions based on a lack of mutual understanding. Such mutual understanding is especially significant in times when complex environmental challenges such as climate change make cross-sectoral responses through multi-actor governance ever more important.

Additionally, bringing to light the personal and professional knowledge and experiences on the basis of which individual actors at the local level manage 'their' landscapes and implement adaptation measures can support these actors in becoming aware of their own taken-for-granted assumptions and attachments that guide their decision-making. Facets of those taken-for-granted assumptions, as this study shows, are questions like '*What* place are we adapting to *which* climate change impacts?', 'What do we seek to protect from harmful impacts?', 'Which adaptation solutions are tolerable, and for what reasons are others not?', as well as 'Is it necessary at all to undertake adaptation activities?'. In the context of a changing climate that is projected to further accelerate cliff erosion and coastal squeeze in many maritime regions (cf. IPCC 2013), understanding the societal values that are firmly grounded in local people-place relationships and that shape debates such as those at Godrevy is becoming more and more important for solving disagreements between actors with contrasting viewpoints on the implementation of physical-material adaptation measures. Bringing to light differing perspectives on the same physical spaces and reflecting on contrasting framings of climate change and erosion makes it possible to engage local stakeholders in open discussions about otherwise unspoken values and foster joint decision-making. A valuable methodological contribution to such process would be participatory stakeholder mapping (cf. e.g. Brown & Raymond 2014, Vergara-Asenjo et al. 2015) that could provide a strong basis for structured dialogue about specific physical-material landscape changes like those planned at Godrevy, and illustrate practically different local viewpoints on land use, access, landscape management, and the meaning of coastal erosion.

8.3 Avenues for Future Research

Based on the empirical data from Cornwall and Godrevy, I highlighted the importance of considering local contexts, people-place relationships, and regional identities in processes of climate change adaptation. I showed how such processes are influenced by place-related factors such as landscape narratives, specific actor constellations, individual perspectives, and local structures of ownership. Nonetheless, this book is just a building block of research on the relevance of people-place relationships for climate adaptation, and many questions remain open. Orienting towards the Place-Based Framework for Local Climate Adaptation presented in section 7.4.3, I propose here avenues for future research regarding actor constellations at the local level, the roles of individual decision-makers, regional identities, local landscape narratives, and the physical-material implementation of suggested adaptation measures.

Regarding local actor constellations and their significance for decision-making about landscape change, future research should address the concrete negotiation processes between such actors in consideration of the underlying power relations, rights and responsibilities, and legal frameworks that shape particular local contexts. What actors and organisations are involved in such negotiations? How can they enforce their interests, if at all? What roles do governmental bodies take on in contrast to non-governmental ones? Can local interests prevail against national agendas? Of particular importance in this context is further research on how to engage local stakeholders in a structured dialogue process where issues of climate change can be addressed in a way that takes into account different narratives and place-based values. A valuable tool for stakeholder engagement was proposed by Gerkensmeier & Ratter (2016); however, empirical research on how to foster such values-based dialogue in practice is still in its early days.

Considering the role of individual decision-makers on the ground, I showed that their perceptions of the landscape and climate adaptation are meldings of their personal experiences and professional backgrounds. In this context, future research should investigate the extent to which these actors' make decisions based on their personal attachments and/or professional training; a question is likely a task for disciplines such as environmental psychology. Questions to address regarding these actors' roles could be: Which 'parts' of their identities do they draw on when making decisions around climate adaptation and landscape management? Are they aware of the tensions under which they stand (cf. section 7.3); and if yes, do they seek to avoid personal bias in their professional roles? The method of walking interviews, possibly in combination with methods from psychological inquiry, would be a valuable means for uncovering in more depth individual actors' relationships with the places and landscapes they manage.

When focusing on the influence of regional identity on processes of climate adaptation, the particular characteristics of such identity, for instance the 'end of the land'-mentality or the Celtic roots referred to, are highly context-specific to Cornwall. Nonetheless, distinct identities of other regions are likely to also influence how climate adaptation is approached locally. Studies about the role of regional identity could be particularly interesting in places with similarly strong such identities, for example Cataluña in Spain, where people also emphasise their otherness and strive for autonomy from their nation state. However, also in regions with less distinct identities, open questions for research are: What elements of collective people-place relationships shape how climate adaptation is addressed locally? Under what circumstances do shared place attachments foster action in response of the changing climate, and when are they a barrier? What elements of such identity can be drawn on to mobilise adaptive actions?

With focus on the distinct landscape narratives, much room is left for research on the details of these societal constructions and the dynamics that arise when different such narratives oppose in decision-making processes. Facets of the narratives that require deeper investigation are, for instance, particular perspectives on the human-nature relationship and the concrete influence of these perspectives on the implementation (or obstruction) of physical adaptation measures. Comparative research in Europe and other locations should assess the extent to which such landscapes narratives vary from place to place, what other societal constructions than presented here can be uncovered, and to which extent such narratives play a role in formal planning and policy processes around climate adaptation. Regarding Cornwall specifically, the question arises what hidden narratives, potentially also with more negative connotations, exist among a wider set of actors and, importantly, the local population. Lastly, with a view to the impacts of landscape narratives on the reshaping of distinct physical spaces, case studies would be useful that retrospectively investigate the negotiation of adaptation measures *after* their implementation, thus reconstructing the underlying constructions of these places and the dynamics of the local actor constellations that resulted in said implementation. The Godrevy case study would be a suitable example in this context, as at the time of my empirical field phase, the start of the actual relocation of the road and car park was scheduled by the National Trust for late 2017.

Closing Remarks

Adger et al. (2011: 20) find that "[...] there are limits to adaptation options for places and their cultures", acknowledging that a society's relationship with their physical environment shapes the array of climate change responses taken into consideration. Although I would not choose such a negative wording, my findings support the conclusion that the approaches to climate change adaptation that different local actors choose are importantly influenced by how they perceive the places in question, how they construct interpret their physical-material components; and what elements of these places they seek to preserve. These interpretations, I argue, set the direction for which adaptation measures are deemed necessary, possible, favourable, or unacceptable. As I highlighted in the opening quote of this study (p. 1), adaptation to climate change therefore is not a technical process taking place in a vacuum of societal contexts, but is always negotiated on the basis of existing ascriptions of meaning to the physical spaces altered by the changing climate. With continuously greater emphasis on multi-agency working to achieve climate change adaptation in landscape management, it is therefore important for future research to investigate the diverse perceptions people have of the places they manage, to secure effective action at the local level.

Uncovering such people-place relationships through the use of a constructivist landscape approach and qualitative, mobile methods anchors debates around climate change and responses to it in local contexts, thus constituting an important tool for comprehending the coming-into-being of adaptation decisions embedded in their respective societal contexts by letting local actors speak to us 'through' the places where they live and which they manage.

References

ADGER, W. N., ARNELL, N. W. & E. L. TOMPKINS (2005): Successful adaptation to climate change across scales. In: Global Environmental Change, Vol. 15, No. 2, pp. 77–86.

ADGER, W. N., BARNETT, J., BROWN, K., MARSHALL, N. & K. O'BRIEN (2013): Cultural dimensions of climate change impacts and adaptation. In: Nature Climate Change, Vol. 3, No. 2, pp. 112–117.

ADGER, W. N., BARNETT, J., CHAPIN III, F. S. & H. ELLEMOR (2011): This Must Be the Place: Underrepresentation of Identity and Meaning in Climate Change Decision-Making. In: Global Environmental Politics, Vol. 11, No. 2, pp. 1–25.

ADGER, W. N., LORENZONI, I. & K. O'BRIEN (eds) (2009): Adapting to Climate Change. Thresholds, Values, Governance. (Cambridge University Press) Cambridge.

AGRAWAL, A. (2010): Local Institutions and Adaptation to Climate Change. In: MEANRS, R. & A. NORTON (eds): Social Dimensions of Climate Change: Equity and Vulnerability in a Warming World. New Frontiers of Social Policy. (World Bank Publications) Washington, D.C., pp. 173–198.

AGRAWAL, A., KONONEN, M. & N. PERRIN (2009): The Role of Local Institutions in Adaptation to Climate Change. World Bank Social Development Working Papers, No. 118, June 2009. Washington, D.C.

AGYEMAN, J., DEVINE-WRIGHT, P. & J. PRANGE (2009): Close to the edge, down by the river? Joining up managed retreat and place attachment in a climate changed world. In: Environment and Planning A, Vol. 41, No. 3, pp. 509–531. Available at: http://www.envplan.com/abstract.cgi?id=a41301. Last requested online on Jan 26, 2015.

ALKON, A. H. (2004): Place, Stories, and Consequences. Heritage Narratives and the Control of Erosion on Lake County, California, Vineyards. In: Organization and Environment, Vol. 17, No. 2, pp. 145–169.

AMUNDSEN, H. (2015): Place attachment as a driver of adaptation in coastal communities in Northern Norway. In: Local Environment: The International Journal of Justice and Sustainability, Vol. 20, No. 3, pp. 257–276. DOI: 10.1080/13549839.2013.838751.

ANDERSON, J. (2004): Talking whilst walking: a geographical archaeology of knowledge. In: Area, Vol. 36, No. 3, pp. 254–261.

ANTROP, M. (2004): Uncertainty in planning metropolitan landscapes. In: TRESS, G., TRESS, B., HARMS, B., SMEETS, P. & A. VAN DER VALK (eds): Planning Metropolitan Landscapes. Concepts, Demands, Approaches (DELTA Series, Vol. 4). (Delta Program, Alterra Green World Research) Wageningen, pp. 12–25.

ANTROP, M. (2013): A brief history of landscape research. In: HOWARD, P., THOMPSON, I. & E. WATERTON (eds): The Routledge Companion to Landscape Studies. (Routledge) Oxon, New York, pp. 12–22.

© Springer Fachmedien Wiesbaden GmbH, part of Springer Nature 2019
V. Köpsel, *New Spaces for Climate Change*, RaumFragen: Stadt –
Region – Landschaft, https://doi.org/10.1007/978-3-658-23313-6

ARBUCKLE, J. G., WRIGHT MORTON, L. & J. HOBBS (2015): Understanding Farmer Perspectives on Climate Change Adaptation and Mitigation: The Roles of Trust in Sources of Climate Information, Climate Change Beliefs, and Perceived Risk. In: Environment and Behavior, Vol. 47, No. 2, pp. 208–234.

BAPTISTE, A. K. (2013): Locals vs. Expert Perception of Climate Change: An Analysis of Fishers in Trinidad and Tobago. In: MUGA, H. & K. THOMAS (eds): Cases on the Diffusion and Adoption of Sustainable Development Practices. (Information Science Reference) Hershey, pp. 44–82.

BARNETT, J. (2010): Adapting to climate change: three key challenges for research and policy–an editorial essay. In: WIREs Climate Change, Vol. 1, No. 3, pp. 314–317.

BAXTER, P. & S. JACK (2008): Qualitative Case Study Methodology: Study Design and Implementation for Novice Researchers. In: The Qualitative Report, Vol. 13, No. 4, pp. 544–559.

BBC PRONOUNCING DICTIONARY OF BRITISH NAMES (1971): Godrevy. (Oxford University Press) Oxford, p. 62.

BECK, U., BONSS, W. & C. LAU (2001): Theorie reflexiver Modernisierung - Fragestellungen, Hypothesen, Forschungsprogramme. In: BECK, U. & W. BONSS (eds): Die Modernisierung der Moderne. (Suhrkamp) Franfurt/Main, pp. 11–62.

BEER, H. (2016): Poldark: behind the scenes. National Trust Magazine, Autumn 2016. Available at: www.nationaltrust.org.uk/features/cornish-coast-stars-in-poldark-remake. Last requested online on Sept 20, 2016.

BIESBROEK, R., SWART, R., CARTER, T., COWAN, C., HENRICHS, T., MELA, H., MORECROFT, M. & D. REY (2010): Europe adapts to climate change: Comparing National Adaptation Strategies. In: Global Environmental Change, Vol. 20, No. 3, pp. 440–450.

BOHNET, I. C. & R. BEILIN (2015): Editorial: Pathways towards sustainable landscapes. In: Sustainability Science, Vol. 10 (Special Issue), pp. 187–194.

BOURS, G., MCGINN, C. & P. PRINGLE (2013): Monitoring & evaluation for climate change adaptation: A synthesis of tools, frameworks and approaches. (Sea Change Community/UKCIP) Phnom Penh, Oxford.

BOWEN, G. A. (2009): Document Analysis as a Qualitative Research Method. In: Qualitative Research Journal, Vol. 9, No. 2, pp. 27–40.

BRACE, C. & H. GEOGHEGAN (2010): Human geographies of climate change: Landscape, temporality, and lay knowledges. In: Progress in Human Geography, Vol. 35, No. 5, pp. 284–302.

BROWN, G. & C. RAYMOND (2014): Methods for identifying land use conflict potential using participatory mapping. In: Landscape and Urban Planning, Vol. 122, pp. 196–208.

BRUNS, D. & O. KÜHNE (2013): Landschaft im Diskurs. Konstruktivistische Landschaftstheorie als Perspektive für künftigen Umgang mit Landschaft. In: Naturschutz und Landschaftsplanung, Vol. 45, No. 3, pp. 83–88.

BRYAN, E., RINGLER, C., OKOBA, B., KOO, J., HERRERO, M. & S. SILVESTRI (2013): Can agriculture support climate change adaptation, greenhouse gas mitigation and rural livelihoods? Insights from Kenya. In: Climatic Change, Vol. 118, No. 2, pp. 151–165.

BRYMAN, A. (2012): Social Research Methods. 4th Edition. (Oxford University Press) Oxford.

BUCIEK, K., BAERENHOLD, O. & K. JUUL (2006): Whose Heritage? Immigration and Place Narratives in Denmark. In: Geografiska Annaler - Series B, Human Geography, Vol. 88, No. 2, pp. 185–197.

BUGLER, W. & O. PALIN (2017): From the ground up: How communities can collaborate to drive local adaptation and influence the national agenda. Climate & Development Knowledge Network Policy Brief. Available at: https://cdkn.org/resource/policy-brief-ground-communities-can-collaborate-drive-local-adaptation-influence-na-tional-agenda/?loclang=en_gb. Last requested online on Apr 26, 2017.

BUNDESREGIERUNG DEUTSCHLAND (2008): Deutsche Anpassungsstrategie an den Klimawandel. Berlin.

BUTLER, J., SUADNYA, W., PUSPADI, K., SUTARYONO, Y., WISE, R., SKEWES, T., KIRONO, D., BOHENSKY, E., HANDAYANI, T., HABIBI, P., KISMAN, M., SUHARTO, I., SUPAR-TARNINGSIH, S., RIPALDI, A., FACHRY, A., YANUARTATI, Y., ABBAS, G., DUGGAN, K. & A. ASH (2014): Framing the application of adaptation pathways for rural livelihoods and global change in eastern Indonesian islands. In: Global Environmental Change, Vol. 28, pp. 368–382.

CADGWITH COTTAGES (2017): The Cot, Cadgwith Cove. Available at: http://www.cadgwith-cottages.co.uk/. Last requested online on Jun 6, 2017.

CAPRIANO, R. (2009): Come take a walk with me: The "Go-Along" interview as a novel method for studying the implications of place for health and well-being. In: Health & Place, Vol. 15, pp. 263–272.

CASTREE, N. (2013): Making sense of nature. Representation, politics and democracy. 2nd Edition. (Routledge) London, New York.

CHARMAZ, K. (2006): Constructing Grounded Theory. A Practical Guide Through Qualitative Analysis. (Sage Publications) Thousand Oaks, London, New Delhi.

CHASE, S. (2005): Narrative Inquiry: Multiple lenses, Approaches, Voices. In: DENZIN, N. K. & Y. S. LINCOLN (eds): The SAGE Handbook of Qualitative Research. (Sage Publications) Thousand Oaks, London, New Delhi, pp. 651–680.

CIOS FUTURES GROUP (2017): A Catalyst for Change. Implications, Risks and Opportunities of Brexit for Cornwall and the Isles of Scilly. Truro. Available at: https://www.cornwall.gov.uk/media/24227365/catalyst-for-change-brexit-re-port.pdf. Last requested online on May 2, 2017.

CLARK, A. & N. EMMEL (2008): Walking interviews: more than walking and talking. Presented at 'Peripatetic Practices': a workshop on walking (London, March 2008).

CLIMATE UK (2012): A Summary of Climate Change Risks for South West England. To coincide with the publication of the UK Climate Change Risk Assessment (CCRA) 2012. Available at: http://climateuk.net/resource/regional-summaries-uk-climate-change-risk-assessment. Last requested online on 26 June 2015.

CLIMATE VISION (2017): About us. Available at: http://climatevision.co.uk/about-us. Last requested online on May 24, 2017.

CLOKE, P. & J. LITTLE (2005): Introduction: Other Countrysides? In: CLOKE, P. & J. LITTLE (eds): Contested Countryside Cultures: Otherness, marginalisation and rurality. 2nd Edition. (Routledge) New York, pp. 1–17.

COAST NETWORK (2017): About us. Available at: http://www.coastproject.co.uk/about. Last requested online on May 24, 2017.

COLLINS, H. & R. EVANS (2002): The Third Wave of Science Studies: Studies of Expertise and Experience. In: Social Studies of Science, Vol. 32, No. 2, pp. 235–296.

COMMISSION OF THE EUROPEAN COMMUNITIES (CEC) (2007): Adapting to climate change in Europe – options for EU action. Green Paper from the Commission to the Council, the European Parliament, the European Economic and Social Committee and the Committee of the Regions. Brussels.

CORBIN, J. (1998): Alternative interpretations: Valid or not? In: Theory & Psychology, Vol. 8, pp. 121–128.

CORNISH GUARDIAN (2015): Severe weather causes travel disruption across Cornwall. Dec 30, 2015. Available at: http://www.cornishguardian.co.uk/Severe-weather-causes-travel-disruption-Cornwall/story-28439656-detail/story.html. Last requested online on Jan 4, 2016.

CORNISH HEDGERS (2017): Introduction to the Cornish Hedge. Available at: http://www.cornishhedgers.org.uk/page.php?nm=intro. Last requested online on Jan 4, 2017.

CORNISH LANGUAGE PARTNERSHIP (2017): Cornish Dictionary - Gerlyver Kernewek: tewyn. Available at: http://www.cornishdictionary.org.uk/browse?field_word_value=tewyn. Last requested online on Feb 27, 2017.

CORNISH MINING WORLD HERITAGE (2017): World Heritage. Available at: http://www.cornish-mining.org.uk/world-heritage. Last requested online on May 24, 2017.

CORNISH NATURE (2010): Grey Seals of the north Cornish coast. Available at: http://www.cornishnature.co.uk/cornish-nature-diary/49-grey-seals-of-the-north-cornish-coast.html. Last requested online on Jan 31, 2017.

CORNWALL AONB PARTNERSHIP (2011a): Cornwall AONB Management Plan 2011-2016. Further Information to each Strategic Chapter: Web-based Appendices. 3. Climate Change and Energy. Truro.

CORNWALL AONB PARTNERSHIP (2011b): Management Plan 2011-2016. Safeguarding our landscape's beauty & benefits for future generations. Truro.

CORNWALL AONB PARTNERSHIP (2016a): 06 - Godrevy to Portreath. Cornwall AONB Official Website. Available at: http://www.cornwall-aonb.gov.uk/godrevytoportreath/. Last requested online on Sept 29, 2016.

CORNWALL AONB PARTNERSHIP (2016b): Cornwall Area of Outstanding Natural Beauty. Offical Website. Available at: http://www.cornwall-aonb.gov.uk/. Last requested online on Dec 19, 2016.

CORNWALL AONB PARTNERSHIP (2017): Planning. Available at: http://www.cornwall-aonb.gov.uk/planning/. Last requested online on Feb 27, 2017.

CORNWALL COMMUNITY FLOOD FORUM (2017): About the Flood Forum. Available at: https://www.cornwall.gov.uk/community-and-living/cornwall-fire-and-rescue-service-homepage/keeping-safe/flooding/cornwall-community-flood-forum/about-the-flood-forum/. Last requested online on May 24, 2017.

CORNWALL COUNCIL (2009): Cornwall Council Local Development Framework. Strategic Flood Risk Assessment Level 1. Truro. Available at: https://www.cornwall.gov.uk/environment-and-planning/planning/planning-policy/cornwall-local-plan/cornwall-topic-based-evidence-base/cornwall-strategic-flood-risk-assessment-%28sfra%29/. Last requested online on Jun 26, 2015.

CORNWALL COUNCIL (2011a): Cornwall Landscape Character: Best Practice Guide Truro. Available at: http://www.cornwall.gov.uk/media/3627266/Landscape_Best_Practice_Aug_2011_Full-version-Web.pdf. Last requested online on 3 March 2016.

CORNWALL COUNCIL (2011b): Preliminary Flood Risk Assessment. Available at: https://www.cornwall.gov.uk/media/6959241/Cornwall-PFRA-Report-June-2011.pdf. Last requested online on May 24, 2017.

CORNWALL COUNCIL (2012): 2011 Census at a Glance. Available at: https://www.cornwall.gov.uk/council-and-democracy/data-and-research/data-by-topic/2011-census/. Last requested online on 14 June 2016.

CORNWALL COUNCIL (2013a): Cornwall Design Guide - Part Two, Truro. Available at: http://www.cornwall.gov.uk/media/13042645/CDG-Sections-1-2-3.pdf. Last requested online on Jan 3, 2017.

CORNWALL COUNCIL (2013b): Technical Paper E2 - An Assessment of the Renewable Energy Resource Potential in Cornwall Truro. Available at: https://www.cornwall.gov.uk/media/3626590/E2-Renwable-Energy-Resource-Potential-_March-2013_.pdf. Last requested online on Jan 9, 2017.

CORNWALL COUNCIL (2014a): Environment and Planning - Cornwall's Landscape. Official Council Website. Available at: http://www.cornwall.gov.uk/environment-and-planning/cornwalls-landscape/. Last requested online on 3 March 2016.

CORNWALL COUNCIL (2014b): What is Cornwall Council's position on climate change? Official Council Website. Available at: https://www.cornwall.gov.uk/environment-and-planning/sustainable-development/climate-change-and-energy/what-is-cornwall-councils-position-on-climate-change/. Last requested online on 3 March 2016.

CORNWALL COUNCIL (2015a): Landscape Character Assessment 2007. Official Council Website. Available at: http://www.cornwall.gov.uk/environment-and-planning/cornwalls-landscape/landscape-character-assessment-2007/. Last requested online on 3 March 2016.

CORNWALL COUNCIL (2015b): What are the potential impacts of climate change in Cornwall? Available at: https://www.cornwall.gov.uk/environment-and-planning/sustainable-development/climate-change-and-energy/what-are-the-potential-impacts-of-climate-change-in-cornwall/. Last requested online on 3 March 2016.

CORNWALL COUNCIL (2016a): Cornwall Council Interactive Map. Available at: https://map.cornwall.gov.uk/website/ccmap/?zoom-level=5&xcoord=154152&ycoord = 40004&wsName=ccmap&layerName=. Last requested online on Sept 26, 2016.

CORNWALL COUNCIL (2016b): Cornwall Local Plan. Strategic Policies 2010-2030 Truro. Available at: https://www.cornwall.gov.uk/media/17155253/local-plan-combined-version-jan-2016-small.pdf. Last requested online on 22 March 2016.

CORNWALL COUNCIL (2016c): Cornwall's Landscape - Think Cornwall, Think Landscape. Available at: https://www.cornwall.gov.uk/environment-and-planning/cornwalls-landscape/. Last requested online on Jan 6, 2017.

CORNWALL COUNCIL (2017a): A30 Temple to Higher Carblake Improvement. Available at: https://www.cornwall.gov.uk/transport-and-streets/roads-highways-and-pavements/major-highway-schemes/a30-temple-to-higher-carblake-improvement/. Last requested online on Jan 6, 2017.

CORNWALL COUNCIL (2017b): Cornwall's Environmental Growth Strategy 2015-2065. Delivering Environmental Growth for a Changing Cornwall. Truro. Available at: http://www.cornwall.gov.uk/environmentalgrowth. Last requested online on May 24, 2017.

CORNWALL COUNCIL (2017c): Data Maps and Infografics. Available at: https://www.cornwall.gov.uk/health-and-social-care/public-health-cornwall/joint-strategic-needs-assessment-jsna/data-maps-and-infographics/#-tab-381389. Last requested online on May 24, 2017.

CORNWALL COUNCIL (2017d): Hedges. Available at: https://www.cornwall.gov.uk/environment-and-planning/trees-hedges-and-woodland/hedges/. Last requested online on Jan 4, 2017.

CORNWALL COUNCIL (2017e): Neighbourhood Planning in Cornwall. Available at: https://www.cornwall.gov.uk/environment-and-planning/planning/neighbourhood-planning/neighbourhood-planning-in-cornwall/#-tab-357841. Last requested online on May 26, 2017.

CORNWALL COUNCIL & HER MAJESTY'S GOVERNMENT (2015): Cornwall Devolution Deal. Available at: https://www.cornwall.gov.uk/media/13409340/20150715-cornwall-devolution-deal-final-reformatted-pdf.pdf. Last requested online on May 12, 2017.

CORNWALL COUNCIL , THE COUNCIL OF THE ISLES OF SCILLY, ENVIRONMENT AGENCY & NATURAL ENGLAND (2010): Cornwall and the Isles of Scilly Shoreline Management Plan. Truro. Available at: www.ciscag.org/smpindex.html. Last requested online on Apr 5, 2017.

CORNWALL GUIDE (2017): Brown Willy on Bodmin Moor. Available at: https://www.cornwalls.co.uk/sites/default/files/photos/brown_willy_bodmin_moor.jpg. Last requested online on Jan 4, 2017.

CORNWALL LIVE (2016): Why I'll never shop at Truro's new Waitrose. Jun 17, 2016. Available at: http://www.cornwalllive.com/why-i-ll-never-shop-at-truro-s-new-waitrose/story-29413835-detail/story.html. Last requested online on Jan 7, 2017.

CORNWALL SEAL GROUP (2017): Official Website. Available at: https://www.cornwallsealgroup.co.uk/. Last requested online on Feb 16, 2017.

CORNWALL WILDLIFE TRUST (2016): Living Landscapes. Available at: http://www.cornwallwildlifetrust.org.uk/livinglandscapes. Last requested online on Dec 14, 2016.

CORNWALL WILDLIFE TRUST (2017a): Official Website. Available at: http://www.cornwallwildlifetrust.org.uk/. Last requested online on Feb 22, 2017.

CORNWALL WILDLIFE TRUST (2017b): What we do. Available at: http://www.cornwallwildlifetrust.org.uk/who-we-are/what-we-do. Last requested online on May 24, 2017.

CORNWALL WILDLIFE TRUST (2018): Cornwall Beaver Project – Offical Website. Available at www.cornwallwildlife.org.uk/beaverproject. Last requested online on May 26, 2018.

COSGROVE, D. & S. DANIELS (eds) (1988): The Iconography of Landscape. (Cambridge University Press) Cambridge.

COUNCIL OF EUROPE (2000): European Landscape Convention. - European Treaty Series, No. 176. Florence.

CRESSWELL, T. (2004): Place: A Short Introduction. (Blackwell) Oxford, Carlton.

CRESSWELL, T. (2009): Place. In: KITCHIN, R. & N. THRIFT (eds): International Encyclopedia of Human Geography, Vol. 8. (Elsevier) Amsterdam, pp. 169–177.

CRESSWELL, T. (2015): Place. An Introduction. 2nd Edition. - Short Introductions to Geography Series. (John Wiley & Sons) Oxford, Chichester.

CRISP, G., PIRRIE, D., SHAIL, R. K. & R. C. SABIN (2001): The Godrevy dog - Early canine or lost pet? In: Geoscience in south-west England, Vol. 10, No. 2, pp. 172–176.

DAILY MAIL ONLINE (2013): Wind farms are a 'complete scam', claims the Environment Secretary who says turbines are causing 'huge unhappiness'. Jun 7, 2013. Available at: http://www.dailymail.co.uk/news/article-2337466/Wind-farms-complete-scam-claims-Environment-Secretary-says-plans-turbines-causing-huge-unhappiness.html. Last requested online on Jul 21, 2017.

DAVY, B. (2009): Parzellen, Allmenden, Zwischenräume - Raumplanung durch Eigentumsgestaltung. In: BERNHARD, C., KILPER, H. & T. MOSS (eds): Im Interesse des Gemeinwohls. Regionale Gemeinschaftsgüter in Geschichte, Politik und Planung. (Campus) Frankfurt, New York, pp. 293–329.

DAWSON, R. J. (ed) (2015): A Climate Change Report Card for Infrastructure. LWEC Report Card. Living With Environmental Change. Available online at http://www.nerc.ac.uk/research/partnerships/ride/lwec/report-cards/infrastructure/. Last requested online on Jul 12, 2017.

DEACON, B. (2010): Cornwall and the Cornish. (Alison Hodge) Penzance.

DEFRA (2004): Making space for water. Developing a new Government strategy for flood and coastal erosion risk management in England. (Defra Publishers) London.

DEMUZERE, M., ORRU, K., HEIDRICH, O., OLAZABAL, E., GENELETTI, D., ORRU, H., BHAVE, A. G., MITTAL, N., FELIU, E. & M. FÄHNLE (2014): Mitigating and adapting to climate change: Multi-functional and multi-scale assessment of green urban infrastructure. In: Journal of Environmental Management, Vol. 146, pp. 107–115.

DENZIN, N. K. & Y. S. LINCOLN (eds) (2005): The SAGE Handbook of Qualitative Research. (Sage Publications) Thousand Oaks, London, New Delhi.

DEPARTMENT FOR COMMUNITIES AND LOCAL GOVERNMENT (2012): National Planning Policy Framework. Available at: https://www.gov.uk/government/uploads/system/uploads/attachment_data/file/6077/2116950.pdf. Last requested online on Feb 27, 2017.

DEPARTMENT FOR COMMUNITIES AND LOCAL GOVERNMENT (2014): Consultation and pre-decision matters. Available at: https://www.gov.uk/guidance/consultation-and-pre-decision-matters#Statutory-consultees-on-applications. Last requested online on Mar 2, 2017.

DESILVEY, C. (2012): Making sense of transience: an anticipatory history. In: Cultural Geographies, Vol. 19, No. 1, pp. 31–54.

DEVINE-WRIGHT, P. (2009): Rethinking NIMBYism: The role of place attachment and place identity in explaining place-protective action. In: Community & Applied Social Psychology, Vol. 19, No. 6, pp. 426–441.

DEVINE-WRIGHT, P. (2013): Think global, act local? The relevance of place attachments and place identities in a climate changed world. In: Global Environmental Change, Vol. 23, No. 1, pp. 61–69.

DEVINE-WRIGHT, P. (2014): Dynamics of Place Attachment in a Climate Changed World. In: MANZO, L. C. & P. DEVINE-WRIGHT (eds): Place Attachment. Advances in Theory, Methods and Applications. (Routledge) Oxford, New York, pp. 165–177.

DEVINE-WRIGHT, P. (2015): Local attachments and identities: a theoretical and empirical project across disciplinary boundaries. In: Progress in Human Geography, Vol. 39, No. 4, pp. 527–530.

DEY, I. (1999): Grounding Grounded Theory. (Academic Press) San Diego.

DÖRING, M. & B. RATTER (2017): The regional framing of climate change: towards a place-based perspective on regional climate change perception in north Frisia. In: Journal of Coastal Conservation. Available at: https://link.springer.com/article/10.1007%2Fs11852-016-0478-0. Last requested online on Jun 19, 2017.

DURKHEIM, E. & M. MAUSS (1963): Primitive Classification. (University of Chicago Press) Chicago.

EARLIE, C. (2015): Field Observations of Wave Induced Coastal Cliff Erosion, Cornwall UK. - PhD Thesis, University of Plymouth. Available at: https://pearl.plymouth.ac.uk/bitstream/handle/10026.1/3526/2015EARLIE775648PHD.pdf?sequence=1. Last requested online on Jan 27, 2017.

EMERICH, M. (2012): A New American Deam: Taking the Celtic Cure in Mediated Landscapes. In: TREGIDGA, G. (ed): Memory, place and identity. The cultural landscapes of Cornwall. (Francis Boutle Publishers) London, pp. 80–89.

ENTMAN, R. (1993): Framing: Toward Clarification of a Fractured Paradigm. In: Journal of Communication, Vol. 43, No. 4, pp. 51–58.

ENVIRONMENT AGENCY (2012a): East Cornwall Catchment Flood Management Plan. Summary Report. Managing Flood Risk. Exeter.

ENVIRONMENT AGENCY (2012b): West Cornwall Catchment Flood Management Plan. Managing Flood Risk. Summary Report. Exeter.

EUROPEAN ENVIRONMENT AGENCY (EEA) (2013): Adaptation in Europe. Addressing risks and opportunities from climate change in the context of socio-economic developments. - EEA report, No. 3/2013 Luxembourg.

EVANS, J. & P. JONES (2011): The walking interview: Methodology, mobility and place. In: Applied Geography, Vol. 31, No. 2, pp. 849–858.

FADER, M., BÖHNER, J. & G. GEROLD (2012): Precipitation Variability and Landscape Degradation in Rio Negro (Argentina). In: Geoöko, Vol. 33, No. 1-2, pp. 5–33.

FALL, J. (2002): Divide and rule: Constructing human boundaries in 'boundless nature'. In: GeoJournal, Vol. 58, No. 4, pp. 243–251.

FAZEY, I., WISE, R., LYON, C., CÂMPEANU, C., MOUG, P. & T. DAVIES (2016): Past and future adaptation pathways. In: Climate and Development, Vol. 8, No. 1. Available at: http://rsa.tandfonline.com/doi/abs/10.1080/17565529.2014.989192. Last requested online on May 30, 2017.

FELDMAN, M., SKÖLDBERG, K., BROWN, R. N. & D. HORNER (2004): Making Sense of Stories: A Rhetorical Approach to Narrative Analysis. In: Journal of Public Administration Research and Theory, Vol. 14, No. 2, pp. 147–170.

FIELD, C., BARROS, V., DOKKEN, D. J., MACH, K. J. & M. MASTRANDREA (2014): Climate Change 2014: Impacts, Adaptation, and Vulnerability Part A: Global and Sectoral Aspects. Contribution of Working Group II to the Fifth Assessment Report of the Intergovernmental Panel on Climate Change. (Cambridge University Press) Cambridge.

FLICK, U. (2004): Design and Process in Qualitative Research. In: FLICK, U., KARDOFF, E. v. & I. STEINKE (eds): A Companion to Qualitative Research. (Sage Publications) Thousand Oaks, London, New Delhi, pp. 146–152.

FLICK, U. (2009): An Introduction to Qualitative Research. 4th Edition. (Sage Publications) Thousand Oaks, London, New Delhi.

FLICK, U. (2014): An Introduction to Qualitative Research. 5th Edition. (Sage Publications) Thousand Oaks, London, New Delhi.

FLICK, U., KARDOFF, E. V. & I. STEINKE (eds) (2004): A Companion to Qualitative Research. (Sage Publications) Thousand Oaks, London, New Delhi.

FONTANA, A. & J. H. FREY (2005): The Interview. From Neutral Stance to Political Involvement. In: DENZIN, N. K. & Y. S. LINCOLN (eds): The SAGE Handbook of Qualitative Research. (Sage Publications) Thousand Oaks, London, New Delhi, pp. 695–727.

FREE IMAGES STOCK PHOTOS (2018): St Agnes – Cornish Mining. Available online license-free at https://de.freeimages.com/photo/st-agnes-1209533. Last requested online on March 28, 2018.

FRESQUE-BAXTER, J. & D. ARMITAGE (2012): Place identity and climate change adaptation: a synthesis and framework for understanding. In: WIREs Climate Change, Vol. 3, No. 3, pp. 251–266.

GAILING, L. (2012a): Dimensions of the Social Construction of Landscapes - Perspectives on New Institutionalism. In: Proceedings of the Latvian Academy of Sciences, Section A Social Sciences and Humanites, Vol. 66 (Special Issue), No. 3, pp. 195–205.

GAILING, L. (2012b): Sektorale Institutionensysteme und die Governance kulturlandschaftlicher Handlungsräume. Eine institutionen- und steuerungstheoretische Perspektive auf die Konstuktion von Kulturlandschaften. In: Raumforschung und Raumordnung, Vol. 70, No. 2, pp. 147–160.

GAILING, L. (2013): Landscape is a commons! In: Lo Squaderno - Explorations in Space and Society, Vol. 30, pp. 17–20.

GAILING, L. (2014): Kulturlandschaftspolitik - Die gesellschaftliche Konstituierung von Kulturlandschaft durch Institutionen und Governance. Planungswissenschaftliche Studien zu Raumordnung und Regionalentwicklung, Vol. 4. (Verlag Dorothea Rohn) Detmold.

GAILING, L. & H. KILPER (2009): Shaping Cultural Landscapes through Regional Govern-ance. In: STRUBELT, W. (ed): German Annual of Spatial Research and Policy (Springer) Heidelberg, Berlin, pp. 1–11.

GAILING, L. & M. LEIBENATH (2015): The social construction of landscapes: Two theoret-ical lenses and their empirical applications. In: Landscape Research, Vol. 40, No. 2, pp. 123–138.

GAILING, L. & A. RÖHRING (2008): Kulturlandschaften als Handlungsräume der Regio-nalent- wicklung. Implikationen des neuen Leitbildes zur Kulturlandschaftsgestal-tung. In: RaumPlanung, Vol. 136, pp. 5–10.

GEBHARD, H., GLASER, R., RADTKE, U. & P. REUBER (2007): Natur und Kultur - eine Neu-bestimmung des Verhältnisses. In: GEBHARD, H., GLASER, R., RADTKE, U. & P. REU-BER (eds): Geographie. Physische Geographie und Humangeographie. (Spektrum Akademischer Verlag) München, pp. 933.

GEOGHEGAN, H. & C. LEYSHON (2012): On climate change and cultural geography: far-ming on the Lizard Peninsula, Cornwall, UK. In: Climatic Change, Vol. 113, No. 1, pp. 55–66.

GEOGHEGAN, H. & C. LEYSHON (2014): Shifting Shores: Managing Challenge and Change on the Lizard Peninsula, Cornwall, UK. In: Landscape Research, Vol. 39, No. 6, pp. 631–646.

GEOGRAPH (2009): National Trust car park entrance, Godrevy. Available at: http://www.geograph.org.uk/photo/1545227. Last requested online on Jan 18, 2017.

GEOGRAPH (2010): Looking like summer: Godrevy beach car park in February. Available at: http://s0.geograph.org.uk/geophotos/01/70/61/1706116_acad86ba.jpg. Last re-quested online on Feb 2, 2017.

GEOGRAPH (2017): Road, Hedge and Pasture. Available at: http://www.geo-graph.org.uk/photo/408009. Last requested online on Jan 4, 2017.

GEORGE, A. & A. BENNETT (2004): Case Studies and Theory Development in the Social Sciences. (MIT Press) Cambridge, London.

GERKENSMEIER, B. & B. RATTER (2016): Multi-risk, multi-scale and multi-stakeholder – the contribution of a bow-tie analysis for risk management in the trilateral Wadden Sea Region. In: Journal of Coastal Conservation. Available at: https://link.springer.com/article/10.1007/s11852-016-0454-8. Last requested online on Jun 19, 2017.

GERRING, J. (2007): Case Study Research. Principles and Practices. (Cambridge Univer-sity Press) Cambridge, New York, Melbourne.

GIDDENS, A. (1988): Die Konstitution der gesellschaft: Grundzüge einer Theorie der Strukturierung. (Campus) Frankfurt, New York.

GIESEKING, J. J., MANGOLD, W., KATZ, C., LOW, S. & S. SAEGERT (2014): Section 8: Land-scape: Nature and Culture. In: GIESEKING, J. J., MANGOLD, W., KATZ, C., LOW, S. & S. SAEGERT (eds): The People, Place, and Space Reader. (Routledge) Oxon, New York. Available at: http://peopleplacespace.org/toc/section-8/. Last requested online on Feb 15, 2017.

GILES, M. & L. CRIPPS (2012): Cliff castles and the Cornish landscape. In: TREGIDGA, G. (ed): Memory, place and identity. The cultural landscapes of Cornwall. (Francis Boutle Publishers) London, pp. 12–21.

GLASER, B. G. & A. L. STRAUSS (1967): The discovery of grounded theory. (Aldine) Chicago.

GOVERNMENT OF THE UNITED KINGDOM (2013): The National Adaptation Programme. Making the country resilient to a changing climate, London. Available at: https://www.gov.uk/government/uploads/system/uploads/attachment_data/file/209866/pb13942-nap-20130701.pdf. Last requested online on Jul 21, 2017.

GREIDER, T. & L. GARKOVICH (1994): Landscapes: The Social Construction of Nature and the Environment. In: Rural Sociology, Vol. 59, No. 1, pp. 1–24. Available at: http://onlinelibrary.wiley.com/doi/10.1111/j.1549-0831.1994.tb00519.x/pdf. Last requested online on Mar 17, 2015.

HAJER, M. (1995): The Politics of Environmental Discourse: Ecological Modernization and the Policy Process. (Calderon Press) Oxford.

HAJER, M. (2003): A frame in the fields: policymaking and the reinvention of politics. In: HAJER, M. & H. WAGENAAR (eds): Deliberative Policy Analysis. Understanding Governance in the Network Society. (Cambridge University Press) Cambridge, Melbourne, Cape Town, pp. 88–112.

HALE, A. (2001): Representing the Cornish - Contesting heritage interpretation in Cornwall. In: Tourist Studies, Vol. 1, No. 2, pp. 185–196.

HARASTA, J. (2012): Tourism and the Cornish Alps. In: TREGIDGA, G. (ed): Memory, place and identity. The cultural landscapes of Cornwall. (Francis Boutle Publishers) London, pp. 90–96.

HARVEY, M. (2007): Citizens in defence of something calles science. In: Science as Culture, Vol. 16, pp. 31–48.

HENDRICKS, C. M. (2005): Participatory storylines and their influence on deliberative forums. In: Policy Sciences, Vol. 38, pp. 1–20.

HOKEMA, D. (2012): Landschaft im Wandel ? Zeitgenössische Landschaftsbegriffe in Wissen- schaft, Planungspraxis und Alltag. Promotionsschrift. Berlin.

HOLSTEIN, J. & J. GUBRIUM (1995): The Active Interview. Qualitative Research Methods, Vol. 37. (Sage Publications) London, Thousand Oaks, New Delhi.

HOOK, S. (2008): Landschaftsveränderungen im südlichen Oberrheingebiet und Schwarzwald.Wahrnehmung kulturtechnischer Maßnahmen seit Beginn der 19. Jahrhunderts. Promotionsschrift. (VDM Verlag Dr. Müller) Saarbrücken.

HOWARD, P. (2013): Perceptual Lenses. In: HOWARD, P., THOMPSON, I. & E. WATERTON (eds): The Routledge Companion to Landscape Studies. (Routledge) London, New York, pp. 43–53.

HOWARD, P., THOMPSON, I. & E. WATERTON (eds) (2013): The Routledge Companion to Landscape Studies. (Routledge) London, New York.

HULME, M. (2008): Geographical work at the boundaries of climate change. In: Transactions of the Institute of British Geographers, Vol. 33, No. 1, pp. 5–11.

HULME, M. (2009): Why We Disagree about Climate Change. Understanding Controversy, Inaction and Opportunity. (Cambridge University Press) Cambridge.

HUNZIKER, M. (2010): Die Bedeutungen der Landschaft für den Menschen: objektive Eigen- schaft der Landschaft oder individuelle Wahrnehmung des Menschen? In: Forum für Wissen 2010, pp. 33–41.

HUNZIKER, M., BUCHECKER, M. & T. HARTIG (2007): Space and Place – Two Aspects of the Human-Landscape Relationship. In: KIENAST, F., WILDI, O. & S. GHOSH (eds): A Changing World. Challenges for Landscape Research. (Springer) Heidelberg, London, New York, pp. 47–62.

HÜPPAUF, B. (2007): Heimat – die Wiederkehr eines verpönten Wortes. Ein Populärmythos im Zeitalter der Globalisierung. In: GEBHARD, G., GEISLER, O. & S. SCHRÖTER (eds): Heimat. Konturen und Konjunkturen eines umstrittenen Konzepts. (transcript) Bielefeld, pp. 109–140.

INGOLD, T. (1993): The Temporality of the Landscape. In: World Archeology, Vol. 25, No. 2 (Conceptions of Time and Ancient Society), pp. 152–174.

INGOLD, T. (2000): The Perception of the Environment. Essays in livelihood, dwelling and skill. (Routledge) London, New York, p. 189-208.

INGOLD, T. (2004): Culture on the ground: the world perceived through the feet. In: Journal of Material Culture, Vol. 9, No. 3, pp. 315–340.

INOX GROUP AND HENRY BOOT DEVELOPMENTS LTD (2014): Landscape and Visual Impact Assessment. Revision A. Truro. Available at: http://stadium4cornwall.co.uk/wp-content/uploads/2014/09/Appendix-6.1-Landscape-and-Visual-Impact-Assessment.pdf. Last requested online on Jan 10, 2017.

IPCC (2013): Climate Change 2013: The Physical Science Basis. Contribution of Working Group I to the Fifth Assessment Report of the Intergovernmental Panel on Climate Change. [STOCKER, T.F., QUIN, D., PLATTNER, G.K., TIGNOR, M., ALLEN, S.K., BOSCHUNG, J., NAUELS, A., XIA, Y., BEX, V. & P.M. MIDGLEY (eds)]. (Cambridge University Press) Cambridge.

JACKSON, J. B. (2005 [1984]): Landschaften. Ein Resümee. In: FRANZEN, B. & S. KREBS (eds): Landschaftstheorie. Texte der Cultural Landscape Studies. (Verlag der Buchhandlung Walther König) Köln, pp. 29–44.

JACOBSEN, J. K. (2007): Use of Landscape Perception Methods in Tourism Studies: A Review of Photo-Based Research Approaches. In: Tourism Geographies, Vol. 9, No. 3, pp. 234–253.

JENNINGS, T. (2009): Exploring the invisibility of local knowledge in decision-making: the Boscastle Harbour flood disaster. In: ADGER, W. N., LORENZONI, I. & K. O'BRIEN (eds): Adapting to Climate Change. Thresholds, Values, Governance (Cambridge University Press) Cambridge, pp. 240–254.

JOHNSTONE, B. (1990): Stories, communities, and place: narratives from middle America. (Indiana University Press) Bloomington.

JUHOLA, S. & L. WESTERHOFF (2011): Challenges of adaptation to climate change across multiple scales: a case study of network governance in two European countries. In: Environmental Science & Policy, Vol. 14, No. 3, pp. 239–247.

KELLER, P. & W. N. ADGER (2000): Theory and practice in assessing vulnerability to climate change nd facilitating adaptation. In: Climatic Change, Vol. 47, No. 4, pp. 325–352.

KEYS, R. (2012): Genius Loci: Landscape, Legend and Locality: The Role of Folklore in the Farming and Fishing Communities of the Rame Peninsula. In: TREGIDGA, G. (ed): Memory, place and identity. The cultural landscapes of Cornwall. (Francis Boutle Publishers) London, pp. 97–112.

KILPER, H. & L. GAILING (2013): Die politische Konstruktion von Kulturlandschaften als kollektive Handlungsräume. Die Dessau-Wörlitzer Kulturlandschaft als Beispiel. In: LEIBENATH, M., HEILAND, S., KILPER, H. & S. TZSCHASCHEL (eds): Wie werden Landschaften gemacht? Sozialwissenschaftliche Perspektiven auf die Konstituierung von Kulturlandschaften. (transcript Verlag) Bielefeld, pp. 169–204.

KÖPSEL, V., WALSH, C. & C. LEYSHON (2016): Landscape Narratives in Practice: Implications for Climate Change Adaptation. In: Geographical Journal. Available at: http://onlinelibrary.wiley.com/doi/10.1111/geoj.12203/abstract. Last requested online on Feb 20,2017.

KÖPSEL, V. & C. WALSH (2018): Coastal landscapes for whom? Adaptation challenges and landscape management in Cornwall. In: Marine Policy, Special Issue. Available at https://www.sciencedirect.com/science/article/pii/S0308597X18304032?via%3Dihub. Last requested online on Jun 13, 2018.

KÜHNE, O. (2008): Distinktion - Macht - Landschaft. Zur sozialen Definition von Landschaft. (VS Verlag für Sozialwissenschaften) Wiesbaden.

KÜHNE, O. (2018): Landschaftstheorie und Landschaftspraxis. Eine Einführung aus sozialkonstruktivistischer Perspektive. 2. Auflage. (Springer) Wiesbaden.

KÜHNE, O. (2014): Wie kommt die Landschaft zurück in die Humangeographie? Plädoyer für eine 'konstruktivistische Landschaftsgeographie'. In: Geographische Zeitschrift, Vol. 102, No. 2, pp. 68–85.

LEIBENATH, M. (2013): Konstruktivistische, interpretative Landschaftsforschung: Prämissen und Perspektiven. In: LEIBENATH, M., HEILAND, S., KILPER, H. & S. TZSCHASCHEL (eds): Wie werden Landschaften gemacht? Sozialwissenschaftliche Perspektiven auf die Konstituierung von Kulturlandschaften. (transcript Verlag) Bielefeld, pp. 7–38.

LEIBENATH, M. (2014): Landschaft im Diskurs: Welche Landschaft? Welcher Diskurs? Praktische Implikationen eines alternativen Entwurfs konstruktivistischer Landschaftsforschung. In: Naturschutz und Landschaftsplanung, Vol. 46, No. 4, pp. 124–129.

LEWIS, J. (2003): Design Issues. In: RITCHIE, J. & J. LEWIS (eds): Qualitative Research Practice. A Guide for Social Science Students and Researchers. (Sage Publications) Thousand Oaks, London, New Delhi, pp. 47–76.

LEYSHON, C. & H. GEOGHEGAN (2012): Anticipatory objects and uncertain imminence: cattle grids, landscape and the presencing of climate change on the Lizard Peninsula, UK. In: Area, Vol. 44, No. 2, pp. 237–244.

LEYSHON, C. & T. WALKER (forthcoming): Framing Risk: Implications for Delivering the Ecological Network Approach in Landscape Management. In: Global Environmental Change.

LINDAHL, K. B., BAKER, S. & C. WALDENSTRÖM (2014): Place Perceptions and Controversies over Forest Management: Exploring a Swedish Example. In: Journal of Environmental Policy & Planning, Vol. 15, pp. 201–223.

LINDNERA, M., FITZGERALDA, J., ZIMMERMANN, N., REYER, C., DELZONE, S., VAN DER MAATEN, E, SCHELHAASI, M.-J., LASCH, P., EGGERSA, J., VAN DER MAATEN-THEUNISSEN, M, SUCKOWC, F., PSOMASB, A., POULTER, B. & B. HANEWINKEL (2014): Climate change and European forests: What do we know, what are the uncertainties, and what are the implications for forest management? In: Journal of Environmental Management, Vol. 146, No. 2, pp. 69–83.

LOPÉZ-MARTINÉZ, F. (2017): Visual landscape preferences in Mediterranean areas and their socio-demographic influences. In: Ecological Engineering, Vol. 104, No. A, pp. 205–215.

MARSDEN, P. (2014): Cornish identity: why Cornwall has always been a separate place. In: The Guardian, Apr 26 2014. Available at: https://www.theguardian.com/uk-news/2014/apr/26/survival-of-cornish-identity-cornwall-separate-place. Last requested online on Jan 9, 2017.

MAXQDA (2016): MaxQDA - Analyse Qualitative Data. Available at: http://www.maxqda.com/?s=coding+tree&post_type=page. Last requested online on Nov 7, 2016.

MCCARTHY, J. J., CANZIANI, O., LEARY, N. A., DOKKEN, D. J. & K. S. WHITE (eds) (2001): Climate Change 2001: Impacts, adaptation and vulnerability. Contribution of Working Group II to the Third Assessment Report of the Intergovernmental Panel on Climate Change. (Cambridge University Press) Cambridge.

MEADOWS, D. H., MEADOWS, D. L. & J. RANDERS (1974): The Limits to Growth. A Report for the Club of Rome's Project on the Predicament of Mankind. (Pan Books) Basingstoke.

MET OFFICE (2014): Winter storms, January to February 2014. Available at: http://www.metoffice.gov.uk/climate/uk/interesting/2014-janwind. Last requested online on May 26, 2017.

MICHEEL, M. (2013): Subjektive Konstruktion von (Kultur-)Landschaft in der Alltagspraxis. In: LEIBENATH, M., HEILAND, S., KILPER, H. & S. TZSCHASCHAL (eds): Wie werden Landschaften gemacht? Sozialwissenschaftliche Perspektiven auf die Konstituierung von Kulturlandschaften. (transcript) Bielefeld.

MITCHELL, J. (2008): What Public Access? Access, Commons and Property Rights. In: Social Legal Studies, Vol. 17, No. 3, pp. 351–367.

MITCHELL, M. & M. EGUDO (2003): A Review of Narrative Methodology. - Defence Science and Technology Organisation (Australia). (DSTO Systems Sciences Laboratory) Edinburgh (AUS).

MITCHELL, W. J. (2002): Introduction. In: MITCHELL, W. J. (ed): Landscape and Power. (University of Chicago Press) Chicago, pp. 1–4.

MOLES, K. (2008): A Walk in Thirdspace: Place, Methods and Walking. In: Sociological Research Online, Vol. 13, No. 4. Available at: http://www.socresonline.org.uk/13/4/2.html. Last requested online on May 19, 2017.

MOSER, S. C. & J. A. EKSTROM (2010): A framework to diagnose barriers to climate change adaptation. In: Proceedings of the National Academy of Sciences (USA), Vol. 107, No. 51, pp. 22026–22031.

MOSS, T. (2014): Spatiality of the commons. In: International Journal of the Commons, Vol. 8, No. 2, pp. 457–471.

MUSGRAVE, R. A., MUSGRAVE, P. B. & L. KULLMER (1994): Die öffentlichen Finanzen in Theorie und Praxis. (UTB) Tübingen.

NATIONAL TRUST (2008): Shifting Shores in the South West - Living with a Changing Coastline. Exeter.

NATIONAL TRUST (2015): Shifting Shores. Playing our part at the coast. Available at: https://www.nationaltrust.org.uk/documents/shifting-shores-report-2015.pdf. Last requested online on Sept 26, 2016.

NATIONAL TRUST (2016a): Annual Report 2015/16. Swindon. Available at: http://www.nationaltrustannualreport.org.uk/. Last requested online on May 24, 2017.

NATIONAL TRUST (2016b): Godrevy. Available at: https://www.nationaltrust.org.uk/godrevy. Last requested online on Sept 19, 2016.

NATIONAL TRUST (2016c): National Trust - Cornwall. Available at: https://www.nationaltrust.org.uk/days-out/regionsouthwest/cornwall. Last requested online on Dec 19, 2016.

NATIONAL TRUST (2017a): About the National Trust. Available at: https://www.nationaltrust.org.uk/features/about-the-national-trust. Last requested online on Feb 15, 2017.

NATIONAL TRUST (2017b): Farm Lettings. Available at: https://www.nationaltrust.org.uk/features/farm-lettings. Last requested online on Mar 2, 2017.

NATIONAL TRUST (2017c): For Ever, For Everyone Appeal. Available at: https://www.nationaltrust.org.uk/features/for-ever-for-everyone-appeal. Last requested online on Jan 6, 2017.

NATIONAL TRUST (2017d): Godrevy to Hell's Mouth walk, near Hayle. Available at: https://www.nationaltrust.org.uk/features/coastal-walks-godrevy-to-hells-mouth-walk-near-hayle. Last requested online on Jan 18, 2017.

NATIONAL TRUST (2017e): Our fragile coast around Godrevy. Available at: https://www.nationaltrust.org.uk/godrevy/features/our-fragile-coast-around-godrevy. Last requested online on Jan 27, 2017.

NATURAL ENGLAND (2008): Position Statement: Climate Change. Available at: http://webarchive.nationalarchives.gov.uk/20101015025248/http://www.naturalengland.org.uk/Images/climatechange-ps_tcm6-14806.pdf. Last requested online on May 24, 2017.

NATURAL ENGLAND (2009): The South West: Cornwall and Devon. National Nature Reserves [brochure]. Available at: http://publications.naturalengland.org.uk/publication/33025. Last requested online on Dec 14, 2016.

NATURAL ENGLAND (2011): Think BIG - How and why landscape-scale conservation benefits wildlife, people and the wider economy (NE309). Available at: http://publications.naturalengland.org.uk/publication/30047. Last requested online on 19 April 2016.

NATURAL ENGLAND (2013): Sites of special scientific interest: managing your land. Available at: https://www.gov.uk/guidance/protected-areas-sites-of-special-scientific-interest. Last requested online on Feb 14, 2017.

NATURAL ENGLAND (2015): Areas of outstanding natural beauty: Natural England's role. Available at: https://www.gov.uk/government/publications/areas-of-outstanding-natural-beauty-natural-englands-role/areas-of-outstanding-natural-beauty-natural-englands-role. Last requested online on May 24, 2017.

NATURAL ENGLAND (2017a): About Us - What We Do, Who We Are. Available at: https://www.gov.uk/government/organisations/natural-england/about. Last requested online on Feb 16, 2017.

NATURAL ENGLAND (2017b): Godrevy Head to St Agnes SSSI. Available at: https://designatedsites.naturalengland.org.uk/SiteDetail.aspx?SiteCode=S1003195&SiteName=G&countyCode=6&responsiblePerson=. Last requested online on Feb 16, 2017.

O'BRIEN, K. (2009): Do values subjectively define the limits to climate change adaptation? In: ADGER, W. N., LORENZONI, I. & K. O'BRIEN (eds): Adapting to Climate Change. Thresholds, Values, Governance (Cambridge University Press) Cambridge, pp. 164–180.

O'BRIEN, K. & J. WOLF (2010): A values-based approach to vulnerability and adaptation to climate change. In: WIREs Climate Change, Vol. 1, No. 2, pp. 232–242.

OLSON, M. (1965): The Logic of Collective Action. Public Goods and the Theory of Groups. (Harvard University Press) Cambridge (MA).

O'RIORDAN, T. (1971): Perspectives on resource management. (Pion) London.

ORLOVE, B. (2009): The past, the present and some possible futures of adaptation. In: ADGER, W. N., LORENZONI, I. & K. O'BRIEN (eds): Adapting to Climate Change. Thresholds, Values, Governance. (Cambridge University Press) Cambridge, pp. 131–163.

OSTROM, E. (1990): Governing the Commons: The Evolution of Institutions for Collective Action. (Cambridge University Press) Cambridge.

OSTROM, E. (2005): Understanding Institutional Diversity. (Princeton University Press) Princeton.

OXFORD DICTIONARY ONLINE (2016): Definiton of 'frame' in English. Available at: https://en.oxforddictionaries.com/definition/frame. Last requested online on Nov 7, 2016.

OXFORD DICTIONARY ONLINE (2017a): Narrative. Available at: https://en.oxforddictionaries.com/definition/narrative. Last requested online on May 19, 2017.

OXFORD DICTIONARY ONLINE (2017b): Society. Available at: https://en.oxforddictionaries.com/definition/society. Last requested online on Mar 14, 2017.

PAASI, A. & K. ZIMMERBAUER (2011): Theory and practice of the region: a contextual analysis of the transformation of Finnish regions. In: Treballs de la Societat Catalana de Geografia, Vol. 71/72, pp. 163–178.

PARLIAMENT OF THE UNITED KINGDOM (2000): Countryside and Rights of Way Act 2000. (The Stationery Office) London.

PARLIAMENT OF THE UNITED KINGDOM (2009): Marine and Coastal Access Act 2009. (The Stationery Office) London.

PARRY, M., CANZIANI, O., PALUTIKOF, J., VAN DER LINDEN, P & C. HANSON (eds) (2007): Climate Change 2007: Impacts, Adaptation and Vulnerability. Contribution of Working Group II to the Fourth Assessment Report of the Intergovernmental Panel on Climate Change. (Cambridge University Press) Cambridge.

PELLING, M. (2010): Adaptation to Climate Change: From Resilience to Transformation. (Routledge) Oxford, New York.

PIECHOCKI, R. (2010): Landschaft. Heimat. Wildnis. Schutz der Natur - aber welcher und warum? (C.H. Beck) München.

PONTEE, N. (2013): Defining coastal squeeze: A discussion. In: Ocean & Coastal Management, Vol. 84, pp. 204–207.

RATTER, B. (2001): Natur, Kultur und Komplexität - Adaptives Umweltmanagement am Niagara Escarpment in Ontario, Kanada. (Springer) Berlin, Heidelberg.

RATTER, B. & M. DÖRING (2015): 'Heimat' as a boundary object? Exploring the potentialities of a boundary object to instigate productive science-stakeholder interaction in North Frisia (Germany). In: Environmental Science & Policy, Vol. 54, pp. 448–455.

RATTER, B. & K. GEE (2012): Heimat - A German concept of regional perception and identity as a basis for coastal management in the Wadden Sea. In: Ocean & Coastal Management, Vol. 68, pp. 127–137.

RAYGORODETSKY, G. (2011): Why Traditional Knowledge Holds the Key to Climate Change. United Nations University Article. Sept 28, 2016. Available at: https://unu.edu/publications/articles/why-traditional-knowledge-holds-the-key-to-climate-change.html.

REED, M. S. (2008): Stakeholder participation for environmental management: A literature review. In: Biological Conservation, Vol. 141, No. 10, pp. 2417–2431.

REIN, M. & D. A. SCHÖN (1993): Reframing policy discourse. In: FISCHER, F. & J. FORESTER (ed): The argumentative turn in policy analysis and planning. (Duke University Press) Durham, pp. 145–166.

RELPH, E. (2008): Sense of place and emerging social and environmental challenges. In: EYLES, J. & A. WILLIAMS (eds): Sense of place, health and quality of life. (Ashgate) Aldershot, pp. 31–44.

REVILL, G. (2016): Engaging climate change: cultural geography and worldly theory. In: Social & Cultural Geography, Vol. 17, No. 6, pp. 803–807.

RICHIE, J., LEWIS, J. & G. ELAM (2003): Designing and Selecting Samples. In: RITCHIE, J. & J. LEWIS (eds): Qualitative Research Practice. A Guide for Social Science Students and Researchers. (Sage Publications) Thousand Oaks, London, New Delhi, pp. 77–108.

RITCHIE, J. (2003): The Applications of Qualitative Methods to Social Research. In: RITCHIE, J. & J. LEWIS (eds): Qualitative Research Practice. A Guide for Social Science Students and Researchers. (Sage Publications) Thousand Oaks, London, New Delhi, pp. 24–46.

RITCHIE, J. & J. LEWIS (eds) (2003): Qualitative Research Practice. A Guide for Social Science Students and Researchers. (Sage Publications) Thousand Oaks, London, New Delhi.

RÖHRING, A. & L. GAILING (2005): Institutional problems and management aspects of shared cultural landscapes:. Conflicts and possible solutions concerning a common good from a social science perspective. - IRS Working Paper No. 33. Erkner.

ROSE, M. & J. WYLIE (2006): Animating Landscape. In: Environment and Planning B: Society and Space, Vol. 24, No. 4, pp. 475–479.

ROYAL AIR FORCE (2017): Stations: RRH Porthreath. Available at: http://www.raf. mod.uk/organisation/stations.cfm?selectStation=9E10C87D-04A3-3648-4EF1B66 180E5ACE4. Last requested online on Jan 3, 2017.

RUIZ-BALLESTEROS, E., VALCUENDE, J. M., QUINTERO, V., CORTES, J. A. & E. RUBIO (2009): Naturalizing the Environment. Perceptual Frames, Senses and Resistance. In: Journal of Material Culture, Vol. 14, No. 2, pp. 147–167.

SANDELL, K. (2005): Access, Tourism and Democracy: A Conceptual Framework and the Non-establishment of a Proposed National Park in Sweden. In: Scandinavian Journal of Hospitality and Tourism, Vol. 5, No. 1, pp. 63–75.

SAYER, A. (2000): Realism and Social Science. (Sage Publications) London, Thousand Oaks, New Delhi.

SCANNELL, L. & R. GIFFORD (2013): Personally Relevant Climate Change: The Role of Place Attachment and Local Versus Global Message Framing in Engagement. In: Environment and Behaviour, Vol. 45, No. 1, pp. 60-85.

SCHEELE, U. & J. OBERDÖRFFER (2011): Transformation der Energiewirtschaft: Zur Raumrelevanz von Klimaschutz und Klimaanpassung. Arbeitsgruppe für regionale Struktur- und Umweltforschung (ARSU), 12. Werkstattbericht. – NORD-WEST2050 (ed), Oldenburg. Available at: www.nordwest2050.de. Last requested online on Nov 5, 2014.

SCHENK, W. (2008): Aktuelle Verständnisse von Kulturlandschaft in der deutschen Raumplanung - ein Zwischenbericht. In: Informationen zur Raumentwicklung, No. 5, pp. 271–278.

SCHERR, S. J., SHAMES, S. & R. FRIEDMAN (2010): Defining Integrated Landscape Management for Policy Makers. ecoagriculture Policy Focus, No. 10. Available at: http://www.un.org/esa/ffd/wp-content/uploads/sites/2/2015/10/IntegratedLandscapeManagementforPolicymakers_Brief_Final_Oct24_2013_smallfile.pdf. Last requested online on Apr 3, 2017.

SCHNELL, R., HILL, P. & E. ESSER (2011): Methoden der Empirischen Sozialforschung. 9. Auflage. (Oldenbourg Verlag) München.

SEMM, K. & H. PALANG (2010): Landscape Accessibility: Spaces for Accessibility or Spaces for Communication? In: Living Reviews in Landscape Research, Vol. 4, No. 4, pp. 5–24.

SHELLER, M. & I. URRY (2006): The new mobilities paradigm. In: Environment and Plannning A, Vol. 38, pp. 207–226.

SHORE SURF SCHOOL (2017): Official Website. Available at: http://www.shoresurf.com/. Last requested online on Feb 16, 2017.

SNAPE, D. & L. SPENCER (2003): The Foundations of Qualitative Research. In: RITCHIE, J. & J. LEWIS (eds): Qualitative Research Practice. A Guide for Social Science Students and Researchers. (Sage Publications) Thousand Oaks, London, New Delhi, pp. 1–23.

SOLIVA, R. (2007): Landscape stories: Using ideal type narratives as a heuristic device in rural studies. In: Journal of Rural Studies, Vol. 23, No. 1, pp. 62–74.

SOLNIT, R. (2001): Wanderlust: A history of walking. (Verso) London.

SOUTH WEST COAST PATH ASSOCIATION (2017a): About the Path. Available at: https://www.southwestcoastpath.org.uk/about-coast-path/. Last requested online on May 24, 2017.

SOUTH WEST COAST PATH ASSOCIATION (2017b): Walk - Godrevy Island & The Knavocks. Available at: https://www.southwestcoastpath.org.uk/walksdb/729/. Last requested online on Jan 31, 2017.

SOUTH WEST REGIONAL ASSEMBLY (2008): The South West Climate Change Action Plan for the South West 2008-2010. Available at: http://www.southwest-ra.gov.uk/media/SWRA/Climate%20Change/Climate_Change_Action_Plan.pdf. Last requested online on May 22, 2017.

SOUTHEND ON SEA BOROUGH COUNCIL (2014): The Nottingham Declaration on Climate Change. Available at: http://www.southend.gov.uk/info/200370/protecting_our_environment_and_emergencies/206/climate_change/11. Last requested online on 23 March 2016.

SPENCER, L., RITCHIE, J. & W. O'CONNOR (2003): Analysis: Practices, Principles and Processes. In: RITCHIE, J. & J. LEWIS (eds): Qualitative Research Practice. A Guide for Social Science Students and Researchers. (Sage Publications) Thousand Oaks, London, New Delhi, pp. 199–218.

ST IVES SOCIETY OF ARTISTS (2017): History. Available at: http://www.stisa.co.uk/history/. Last requested online on May 24, 2017.

STAR, S. & J. GRIESEMER (1989): Insitutional Ecology, 'Translations' and Boundary Objects: Amateurs and Professionals in Berkeley's Museum of Vertebrate Zoology, 1907-39. In: Social Studies of Science, Vol. 19, No. 3, pp. 387–420.

STEDMAN, R. (2002): Toward a Social Psychology of Place. Predicting Behavior From Place-Based Cognitions, Attitude, and Identity. In: Environment and Behavior, Vol. 34, No. 5, pp. 561–581.

STRAUSS, A. L. (1987): Qualitative Analysis for Social Scientists. (Cambridge University Press) New York.

SWALLOW, G. (2012): Cribzilla: The Construction of a Cornish Surfing Myth. In: TREGIDGA, G. (ed): Memory, place and identity. The cultural landscapes of Cornwall. (Francis Boutle Publishers) London, pp. 124–140.

TAYLOR, K. (2008): Landscape and Memory: cultural landscapes, intangible values and some thoughts on Asia. In: 16th ICOMOS General Assembly and International Symposium: 'Finding the spirit of place – between the tangible and the intangible', 29 sept – 4 oct 2008, Quebec, Canada. Available at: http://openarchive.icomos.org/139/. Last requested online on Mar 14, 2017.

TAYLOR, A.L., DESSAI, S. & W.B. DE BRUINE (2014): Public perception of climate risk and adaptation in the UK: A review of literature. In: Climate Risk Management, Vol. 4-5, pp. 1-16.

THE GUARDIAN (2016): Cornwall fears loss of funding after backing Brexit. 26 Jun, 2016. Available at: https://www.theguardian.com/uk-news/2016/jun/26/cornwall-fears-loss-of-funding-after-backing-brexit. Last requested online on Jan 6, 2017.

THOMAS, C. & J. MANN (2009): Godrevy Light. (Twelveheads Press) Truro.

THOMPSON, I., HOWARD, P. & E. WATERTON (2013): INTRODUCTION. IN: HOWARD, P., THOMPSON, I. & E. WATERTON (eds): The Routledge Companion to Landscape Studies. (Routledge) London, New York, pp. 1–7.

TOMANEY, J. (2015): Understanding parochialism: a response to Patrick Devine-Wright. In: Progress in Human Geography, Vol. 39, No. 4, pp. 531–532.

TOWANS PARTNERSHIP (2014): Hayle to Godrevy Towans Management Plan. Report by Martin Rule to the Towans Partnership. Available at: http://www.towanspartnership.org.uk/documents/FinalTowansPlan2014.pdf. Last requested online on Jun 22, 2016.

TOWANS PARTNERSHIP (2016a): About Us. Available at: http://www.towanspartnership.org.uk/about.html. Last requested online on Feb 16, 2017.

TOWANS PARTNERSHIP (2016b): St Gothian Sands Local Nature Reserve (On-Site Information Board). Gwithian.

TREGIDGA, G. (ed) (2012): Memory, place and identity. The cultural landscapes of Cornwall. (Francis Boutle Publishers) London.

TRENTMANN, N. (2014): Die Wahrheit hinter der Rosamunde-Pilcher-Fassade. In: Die WELT Online, 03.10.2014. Available at: https://www.welt.de/wirtschaft/article132886150/Die-Wahrheit-hinter-der-Rosamunde-Pilcher-Fassade.html. Last requested online on May 2, 2017.

TUAN, Y. (1977): Space and Place. The Perspective of Experience. (University of Minnesota Press) Minneapolis.

TUAN, Y.-F. (1974): Topophilia: A Study of Environmental Perception, Attitudes and Values. (Columbia University Press) New York.

TURNER, T. (1996): City as a Landscape. A Post-Postmodern View of Design and Planning. (Taylor & Francis) London.

UK NATIONAL PARKS (2017): History of the National Parks. Available at: http://www.nationalparks.gov.uk/students/whatisanationalpark/history. Last requested online on May 18, 2017.

UNESCO (2017): Guide 6 – Case Study: Cornwall and West Devon Mining Landscape (United Kingdom). Available at: http://whc.unesco.org/sustainabletourismtoolkit/guide-6-%E2%80%93-case-study-cornwall-and-west-devon-mining-landscape-united-kingdom. Last requested online on May 26, 2017.

UPHOFF, N. & L. E. BUCK (2006): Strengthening rural local institutional capacities for sustainable livelihoods and equitable development. Paper prepared for the Social Development Department of the World Bank. Washington, D.C.

UPSTREAM THINKING (2017): Official Website. Available at: http://www.upstreamthinking.org/index.cfm?articleid=8692. Last requested online on May 24, 2017.

VAN EETEN, M. (1997): Sprookjes in Rivierenland: Beleidsverhalen over Wateroverlast en Dijkversterking (Fairy tales in Riverland: Policy stories about floods and dike improvement). In: Beleid en Maatschappij, Vol. 24, No. 1, pp. 32–43.

VANDERHEYDEN, V., VAN DER HORST, D, VAN ROMPAEY, A. & S. SCHMITZ (2014): Perceiving the Ordinary: A Study of Everyday Landscapes in Belgium. In: Tijdschrift voor Economische en Sociale Geografie, Vol. 105, No. 2, pp. 591–603.

VERDUIJN, S., MEIJERINK, S. & P. LEROY (2012): How the Second Delta Committee Set the Agenda for Climate Adaptation Policy: A Dutch Case Study on Framing Strategies for Policy Change. In: Water Alternatives, Vol. 5, No. 2, pp. 469–484.

VERGARA-ASENJO, G., SHARMA, D. & C. POTVIN (2015): Engaging Stakeholders: Assessing Accuracy of Participatory Mapping of Land Cover in Panama. In: Conservation Letters, Vol. 8, No. 6, pp. 432–439.

VISIT CORNWALL (2013): Cornwall Visitor Survey 2012. Final Report. Available at: http://www.visitcornwall.com/sites/default/files/generic_files/Cornwall%20Visitor%20Survey%20. Last requested online on May 22, 2017.

VISIT CORNWALL (2017a): Godrevy Beach. Available at: https://www.visitcornwall.com/beaches/good-beach-guide-2014/west-cornwall/hayle/godrevy-beach. Last requested online on Jan 16, 2017.

Visit Cornwall (2017b): World Heritage Site – Discover Mining Heritage. Available at: http://www.visitcornwall.com/things-to-do/history-and-heritage/world-heritage-site. Last requested online on Jul 21, 2017.

WAGNER, J.-M. (1997): Zur emotionalen Wirksamkeit von Kulturlandschaft. In: SCHENK, W., FEHN, K. & D. DENECKE (eds): Kulturlandschaftspflege. Beiträge der Geographie zur räumlichen Planung. (Borntraeger) Stuttgart, pp. 59–66.

WALSH, C. (2010): Strategic Spatial Planning at the Regional and Local Scales: A Case Study of the Dublin City-Region. Available at: https://www.researchgate.net/publication/256461693_Strategic_Spatial_Planning_at_the_Regional_and_Local_Scales_A_Case_Study_of_the_Dublin_City-Region. Last requested online on Sept 2, 2016.

WALSH, C. & S. ALLIN (2012): Strategic Spatial Planning: Responding to Diverse Territorial Development Challenges: Towards an Inductive Comparative Approach. In: International Planning Studies, Vol. 14, No. 4, pp. 377–395.

WARD THOMPSON, C. (2013): Landscape Perception and Environmental Psychology. In: HOWARD, P., THOMPSON, I. & E. WATERTON (eds): The Routledge Companion to Landscape Studies. (Routledge) London, New York, pp. 25–42.

WATTCHOW, B. (2013): Landscape and a sense of place: a creative tension. In: HOWARD, P., THOMPSON, I. & E. WATERTON (eds): The Routledge Companion to Landscape Studies. (Routledge) Oxon, New York, pp. 87–96.

WEBER, E. & P. C. STERN (2011): Public understanding of climate change in the United States. In: American Psychologist, Vol. 66, No. 4, pp. 315–328.

WEHBERG, J., BOCK, M., WEINZIERL, T., CONRAD, O., BÖHNER, J., STELLMES, M. & L. LANDSCHREIBER (2013): Terrain-based Landscape Structure Classification in Relation to Remote Sensing Products and Soil Data for the Okavango Catchment. In: Biodiversity & Ecology, Vol. 5, pp. 221–233.

WHILE, A. & M. SHORT (2011): Place narratives and heritage management: the modernist legacy in Manchester. In: Area, Vol. 43, No. 1, pp. 4–13.

WILBANKS, T., LEIBY, P., PERLACK, R., ENSMINGER, J. T. & S. B. WRIGHT (2007): Toward an integrated analysis of mitigation and adaptation: some preliminary findings. In: Mitigation and Adaptation Strategies for Global Change, Vol. 12, No. 5, pp. 713–725.

WILSON, E. (2007): Adapting to Climate Change at the Local Level: The Spatial Planning Response. In: Local Environment: The International Journal of Justice and Sustainability, Vol. 11, No. 6, pp. 609–625.

WOLF, J., ALLICE, I. & T. BELL (2013): Values, climate change, and implications for adaptation: Evidence from two communities in Labrador, Canada. In: Global Environmental Change, Vol. 23, No. 2, pp. 548–562.

WOODCOCK, P. (2015): Cornwall and Yorkshire show regional identities run deep in England, too. Available at: http://theconversation.com/cornwall-and-yorkshire-show-regional-identities-run-deep-in-england-too-41322. Last requested online on Jun 8, 2017.

WORLD HEALTH ORGANIZATION (2016): Urban green spaces and health. A review of evidence. (WHO Regional Office for Europe) Copenhagen. Available at: http://www.euro.who.int/__data/assets/pdf_file/0005/321971/Urban-green-spaces-and-health-review-evidence.pdf?ua=1. Last requested online on Jan 10, 2017.

WYLIE, J. (2007): Landscape. (Routledge) Oxon, New York.

WYLIE, J. (2009): Landscape, absence and the geographies of love. In: Transactions of the Institute of British Geographers, Vol. 34, No. 3, pp. 275–289.

WYLIE, J. (2011): Landscape. In: AGNEW, J. & D. LIVINGSTONE (eds): The SAGE Handbook of Geographical Knowledge. (Sage Publications) London, Thousand Oaks, New Delhi, pp. 300–315.

WYNNE, B. (2001): Creating public alienation: Expert cultures of risk and ethics on GMOs. In: Science as Culture, Vol. 10, pp. 445–481.

YIN, R. K. (2009): Case Study Research. Design and Methods - Applied Social Research Methods Series, Volume 5. (Sage Publications) Thousand Oaks, London, New Delhi.

YIN, R. K. (2012): Applications of Case Study Research. (Sage Publications) Thousand Oaks, London, New Delhi.

YOUNG, E. (2001): State Intervention and Abuse of the Commons: Fisheries Development in Baja California Sur, Mexico. In: Annals of the Association of American Geographers, Vol. 91, No. 2, pp. 283–306.

YOUNG, I. (1996): Communication and the other: Beyond deliberative democracy. In: BENHABIB, S. (ed): Democracy and Difference. (Princeton University Press) Princeton, pp. 120–135.

Appendices

Appendix 1: Impression of Cornwall's Landscapes

1) Farmland and renewable energy development near Ladock, Cornwall
Source: own photo, 2016

2) Historic harbour of Mousehole, West Penwith, Cornwall
Source: own photo, 2016

3) Historic harbour of Mousehole, West Penwith, Cornwall
Source: own photo, 2016

4) Coastal walk in southwest Cornwall
Source: own photo, 2016

5) Historic centre of Penryn, Cornwall
Source: own photo, 2015

6) Coastal formation at Kynance Cove
Source: own photo, 2016

© Springer Fachmedien Wiesbaden GmbH, part of Springer Nature 2019
V. Köpsel, *New Spaces for Climate Change*, RaumFragen: Stadt – Region – Landschaft, https://doi.org/10.1007/978-3-658-23313-6

Appendix 2: List of analysed documents (alphabetical)

Organisation	Title	Published
Cornwall AONB Partnership	Cornwall AONB Management Plan 2011-2016	2016
	Cornwall AONB Official Website (webpage)	2017
	Godrevy to Portreath AONB (webpage)	2017
Cornwall Council	Cornwall Landscape Character: Best Practice Guide	2011
	Cornwall's Landscape (website)	2014
	Landscape Character Assessment (report)	2015
	Cornwall Local Plan. Strategic Policies 2010-2030	2016
	Think Cornwall, Think Landscape (webpage)	2016
	Environmental Growth Strategy 2015-2065	2017
Cornwall Wildlife Trust	Living Landscapes	2016
	Official Website (webpage)	2017
Environm. Agency	West Cornwall Catchment Flood Management Plan	2012
National Trust	Shifting Shores in the South West - Living with a Changing Coastline	2008
	National Trust – Cornwall (webpage)	2016
	Godrevy to Hell's Mouth Walk (webpage)	2017
	Our fragile coast at Godrevy (webpage)	2017
Natural England	The South West: Cornwall and Devon. National Nature Reserves (brochure)	2009
	Think BIG. How and why landscape-scale conserva-tion benefits wildlife, people and the wider economy	2011
	Sites of special scientific interest: managing your land	2013
	Godrevy Head to St Agnes SSSI (webpage)	2017
SW Coast Path Association	About the Path (webpage)	2017
	Walk - Godrevy Island & The Knavocks (webpage)	2017
Towans Partnership	Hayle to Godrevy Towans Management Plan	2014
	Official Website	2016
	St Gothian Sands Local Nature Reserve (on-site information board)	2016
Visit Cornwall	Godrevy Beach (webpage)	2017

Appendix 3: Overview of Interviewees (field phases 1 and 2)

#	Organisation	Role	Date/Time	Acronym
Field phase 1 – Sept-Nov 2015 – mid-Cornwall				
1	Natural England	Lead Advisor	15.09.2015, 10:00 – 11:50	I-1
2	Community Flood Forum	Head	16.09.2015, 12:30 – 14:00	I-2
3	Parish Councillor	Ladock Parish Council	23.09.2015, 13:00 – 14:15	I-3
4	Cornwall Councillor	Non-aligned	24.09.2015 12:45 – 14:35	I-4
5	Visit Cornwall	Director of Tourism Board	02.10.2015, 17:00 – 18:50	I-5
6	Cornwall Council	Strategic Energy Development	05.10.2015, 16:15 – 17:20	I-6
7	CoaST	Founder/Director	06.10.2015, 18:00 – 19:40	I-7
8	Cornwall Councillor	Independent, formerly Conservative	07.0.2015, 12:15 – 13:15	I-8
9	Cornwall Council	Natural Environment Officer	08.10.2015, 14:15 – 15:00	I-9
10	Parish Councillor	Perranwell Parish Council	12.10.2015, 13:00 – 14:30	I-10
11	Cornwall Wildlife Trust	Executive Director	12.10.2015, 16:15 – 17:30	I-11
12	Environment Agency	Integrated Environmental Planning	13.10.2015, 15:00 – 16:50	I-12
13	National Trust	General Manager (The Lizard)	14.10.2015, 17:00 – 18:30	I-13
14	Climate Vision	Founder/Director	16.10.2015, 15:00 – 16:15	I-14
15	Upstream Thinking	'Upstream Thinking' Project Manager	29.10.2015, 9:30 – 10:40	I-15

16	National Trust	General Manager (mid-Cornwall)	04.11.2015, 13:00 – 14:20	I-16
17	Environment Agency	Flood Risk Manager (Cornwall, Devon)	09.11.2015, 10:00 – 11:10	I-17
18	AONB Partnership	Partnership Manager	12.11.2015, 15:30 – 16:50	I-18
19	Cornwall Council	Strategic Environment Manager	27.11.2015, 10:00 – 11:05	
Field phase 2 – May-Jun 2016 – Godrevy Headland				
1	Towans Partnership	Chair	05.05.2016, 10:00 – 11:25	I-G1.1
	Towans Partnership	Local Ranger	"	I-G1.2
2	Cornwall Wildlife Trust	West Cornwall Reserves Manager	05.05.2016, 11:30 – 12:30	I-G2
3	National Trust	Lead Ranger, Godrevy	10.05.2016, 10:00 – 11:30	I-G3
4	Natural England	Lead Advisor West Cornwall	12.05.2016, 10:15 – 11:35	I-G4.1
	Natural England	Lead Advisor Cornwall	"	I-G4.2
5	National Trust	General Manager (mid-Cornwall)	15.05.2016, 14:00 – 15:20	I-G5
6	Tenant Farmer		18.05.2016, 10:00 – 10:50	I-G6
7	Visit Cornwall	Chief Executive	19.05.2016, 15:00 – 15:45	I-G7
8	Local Resident	Gwinnear-Gwithian Parish Council	24.05.2016, 10:00 – 11:00	I-G8
9	Cornwall Seal Group	Lead Researcher	25.05.2016, 15:00 – 16:15	I-G9
10	Coast Path Association	Planning & Environmental Advisor	31.05.2016, 16:00 – 17:00	I-G10
11	AONB Partnership	Partnership Manager	02.06.2016, 13:30 – 14:30	I-G11
12	Local Surf School	Owner	03.06.2016, 15:00 – 16:00	I-G12

Appendix 4: Interview guide – field phase 1 (mid-Cornwall)

1. *Introduction:*

–   Introduction of researcher
–   Where here in Cornwall do you work and live? Since when?
–   Originally from here? (If not, where?)

2. *Perception of Landscape/Heritage:*

So if you take a second now and think about the landscape here in the region...

–   Now imagine I had never been to Cornwall. How would you describe Cornwall to me?
–   Characteristics? Adjectives?
–   Any negative associations with the landscape?
–   Places here that you feel particularly attached to regarding the landscape?
–   What role does landscape play in Cornishness/sense of place?
–   Regarding job, how does it connect with the landscape?

3. *Landscape Change:*

–   How did landscape here change over the past 10-20 years? Examples? Why?
–   What do you think about these changes?
–   Who changes the landscape in this area? You too?
–   Frequency of flooding is increasing here in the region → Impacts on the landscape?
–   What elements of the landscape would you miss if they were gone, and why?

4. *Climate Change:*

–   Do you experience that the climate is changing here in the region? If so, how?
–   In what everyday activities do you have to deal with it?
–   Professionally, how much of a topic is climate change in your work?
–   And privately, engaged in any activities around climate change and/or adaptation?

5. *Adaptation & Landscape Change:*

–   Coming back to the abovementioned changes in the landscape - are there any changes which you think are due to climate change?
–   Do you know of any adaptive activities already undertaken around here? Who was involved? Negotiation? Opinion?

- Who are the actors involved in adaptation in the region? Who *should* be involved?
- Are you involved in any adaptation activities, yourself?
- Besides current activities, what other landscape changes do you regard necessary to adapt to climate change?

6. *Adaptation measures:*

- When planning adaptation measures, is their visual impact on the landscape an issue for [*your organisation*] in the planning process?
- Would you say [*your organisation*]'s approach to managing the landscape in Cornwall has changed since climate change is on the agenda? How?
- How do you deal with the uncertainty of CC evidence in your work?
- When adaptation measures are being discussed, does [*your organisation*] have possibilities to put their interests regarding implementation through?
- Do you see any conflict with other organizations' opinion on how to adapt to climate change?
- Changing perspective: Do you think climate change and adaptation are issues that the local population in Cornwall is concerned with?

7. *Regional Identity + Adaptation* → Attention back to Cornwall in general

- Is there anything typically Cornish that you would say makes it different how adaptation is approached here than in other parts of England?
- Would you say that there is anything about Cornwall as a place / Cornish culture which makes climate change adaptation particularly hard or easy?

8. *Landscape - Outlook:*

- Let us take a little look into the future. Thinking twenty or so years into the future, how do you imagine the landscape look different from now?
- What is lacking in the current approach to managing the landscape?

Any recommendations for further interviewees?

Appendix 5: Interview guide – field phase 2 (Godrevy)

Thank you for coming out and meeting me here! I think actually being in the place, we can have a very interesting conversation about this part of the landscape and how it is changing. Let's start walking in a direction of your choice, and while we walk…

1. Landscape
- Would you like to start by telling me about the Godrevy landscape here?
- How would you describe your relationship to the landscape in Godrevy?
- To what extent is Godrevy representative of the Cornish landscape?
- Would you say that Godrevy has any wider importance throughout Cornwall other than being appreciated by the locals here?
- Since when has this site been managed by the National Trust/Natural England/ Towans Partnership/…? *(experts)*

2. Change
- What changes have you experienced in the landscape here? How do you view those changes?
- What do you see as the key factors leading to change in the landscape?
- (When the coast erodes like it does here, what does that mean to you?)*

3. Landscape + Climate Change
- Do you see the changes here at the coast in relation to climate change?
- (How) Does climate change impact on your work in managing the landscape here?
- To what extent are local residents and visitors aware of climate change as a factor influencing the landscape? *(experts)*

4. Landscape Management

EXPERTS:
- When managing for change here, what are your priorities? What is most important?
- Are there any particular management challenges you are facing here?
- Who is involved in Godrevy? Whom are you working with?
- (When coastal erosion proceeds, as you say, the road here and the car park will be affected. From an AONB/Trust/NE perspective, how should this be dealt with?)*
- Do you need a different kind of planning from before to be able to adapt to the impacts of climate change?
- What is the time frame we're talking about when it comes to managing issues like this one?

- What role do the visitors and their expectations play when it comes to landscape management here? Do visitor expectations influence management decision-making?
- Does the National Trust have the obligation to ensure access to Godrevy headland?
- When considering relocating the car park, to what extent are alternative options of access to the headland part of the discussion?
  - ⇥ Do you see a potential here to encourage visitors to use other means of transport?

LOCALS:
- When the coast is eroding further here, what do you expect of the National Trust and the other organizations in terms of measures/management?
- The NT want to relocate the road/car park here, what do you think about this?
- (Why) Is it important to you that Godrevy will be accessible in the future?
  - ⇥ Generally in terms of access to the landscape?
  - ⇥ Accessibility by car?

ADDITIONAL QUESTIONS FOR FARMER:
- How long has this land been farmed by you/your family, leased from the Trust?
- As you are National Trust tenants, how much of an influence does the Trust have on your way of managing the land?
- How does the Trust involve you in decision-making regarding the management of the headland?

5. Negotiation Process

EXPERTS:
- I understand that Godrevy is National Trust property. What processes are in place to allow for local participation in decision-making on the future of the landscape?
- In managing this piece of land, to which guidelines are you bound? Or deciding freely?
- Do you see specific challenges in coordinating the decision-making of the National Trust/AONB/Natural England with the other agencies who are involved here?
- To what extent do you perceive potential for conflict in the future?
- Regarding the relocation activities, what options does your organization have to ensure that its interests are followed through?
- In what ways do you communicate your management decisions to local people/visitors?

LOCALS:
- Is the management of the Godrevy area a topic among people who live around here?
- To what extent are you aware of the plans of the National Trust for Godrevy?
- Do you know of any conflicts around the management of Godrevy under climate change?
- And would you say there are differing opinions about this among the local community?

*Will ask this question depending on the content of the conversation so far

Appendix 6: Post-interview notes (walking interviews at Godrevy)

TOWANS PARTNERSHIP – 05.05.2016 – 10:00am

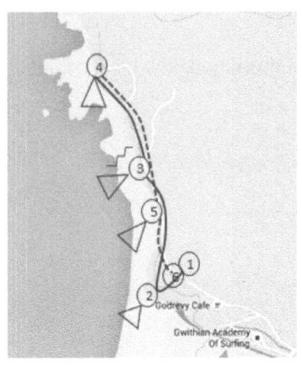

ROUTE:
From main car park to platform overlooking the beach + the Towans, stood there for about 15mins while talking. Then walked across the car park to footpath and towards Godrevy headland. On towards upper car park along the coastline, viewing the erosion and looking at some stairs which had fallen down in the storms. Through to upper car park, looked at lighthouse for a bit. Slowly back to main car park.

2 persons interviewed: Voluntary chairman of Towans Partnership & Towans Ranger

Weather:
- Bright sunshine, barely any clouds, about 17°C; rather windy along the coastline
- Quite crowded, many surfers, low tide with waves

Thoughts:
- Conservationist view, but like natural processes and changes and coastal erosion
- CC usually not a large topic in their work
- Not directly involved with Godrevy, more like neighbours
- NT General Mananger talked to the Towans Partnership about his plans last week
- Cornwall is a place where people agree on the beauty of the landscapes, attachment is high,
- Even though there is agreement about what makes up the Cornish landscapes visually and the relevance of landscape for the region, framings differ a lot when taking a closer look at the organizations in environmental management; important to uncover the reasoning and underlying assumptions behind adaptation approaches
- "Landscape", precisely because it is such a fluffy term, is a good means (as Brace & Geoghegan 2010 found) to 'ground' the climate change discussion locally; also regarding adaptation to its impacts within environmental management organizations
  → There is a lot of vested interest in landscape issues (which tunes in with place attachment stuff and identity)! Talking about the landscape opens up a discussion around many wider topics related to cc adaptation

Appendix 7: Overview of coding in interview analyses

Field phase 1 – broader Cornwall area

Theme	Initial Code
Landscape Perception	What is landscape? Characteristics Activities in the landscape Negative associations How job relates to landscape Landscape knowledge
Landscape Change	Change per se Built development Changes in farming practice Infrastructure development Loss of distinctiveness Future landscape
Attachments	Attachment to Cornwall/appreciation Landscape attachment Attachment to specific places
Climate change + landscape	Perception of climate change Changing landscape under climate change Climate change and job More rainfall → flooding Extreme events Coastal erosion/coastal flooding Different species and diseases Renewables Farming practice
New requirements to landscape management	Sustainability needed Flooding inland Coastal flooding and erosion Changing farming practice Infrastructure Renewables Species change
Adaptation activities	Actors in adaptation Negotiation position/responsibilities General approach to adaptation Concrete measures and examples What's lacking? Clashes/conflicts

Field phase 2 – Godrevy

Theme	Initial Code
Landscape Perception	What is the Godrevy landscape? Significance of Godrevy Aesthetics Relationship with landscape Wording Management activities + priorities Perception of change
Climate change + coastal erosion	Perception of climate change Climate change at Godrevy Perception of coastal erosion Communicating climate change
Access	Access for whom? Beach access People's perspective Management challenges around access
Temporal scales	Management time scale Long-term vs. short-term thinking Visitor concerns
Spatial scales	Spatial scale of management Policies in the background Issues dealt with
Negotiating the landscape	Involved parties Rights + responsibilities User conflicts Perception of the National Trust Management challenges Participation
Making physical space	Climate change changing the space Responses to physical cc impacts New car park + road Beach access

Appendix 8: List of interviewees and narratives they follow

Field phase 1 – broader Cornwall area

#	Organisation	Primary Narrative	Secondary Narrative
1	Natural England	Natural Systems	Human-Environment
2	Community Flood Forum	Human-Environment	Visual Beauty
3	Parish Councillor	Functional Landscape	Natural Systems
4	Cornwall Councillor	Functional Landscape	Human-Environment
5	Visit Cornwall	Visual Beauty	Human-Environment
6	Cornwall Council	Human-Environment	Natural Systems
7	CoaST*	Natural Systems	Human-Environment
8	Cornwall Councillor	Visual Beauty	--
9	Cornwall Council	Natural Systems	Human-Environment
10	Parish Councillor	Natural Systems	Human-Environment
11	Wildlife Trust	Natural Systems	Human-Environment
12	Environment Agency	Natural Systems	Functional Landscape
13	National Trust	Human-Environment	Visual Beauty
14	Climate Vision	Natural Systems	Human-Environment
15	Upstream Thinking	Natural Systems	Functional Landscape
16	National Trust	Human-Environment	Visual Beauty
17	Environment Agency	Human-Environment	Visual Beauty
18	AONB Partnership	Human-Environment	Visual Beauty
19	Cornwall Council	Human-Environment	Visual Beauty

Field phase 2 – Godrevy

#	Organisation	Primary Narrative	Secondary narrative
G-1.1	Towans Partnership	Natural Systems	Natural Beauty
G-1.2	Towans Partnership	Natural Systems	Natural Beauty
G-2	Wildlife Trust	Natural Systems	Natural Beauty
G-3	National Trust	Human-Environment	Natural Systems
G-4.1	Natural England	Natural Systems	Human-Environment
G-4.2	Natural England	Natural Systems	Human-Environment
G-5	National Trust	Human-Environment	Natural Beauty
G-6	Tenant Farmer	Functional	Natural Systems
G-7	Visit Cornwall	Human-Environment	Natural Beauty
G-8	Local Resident	Human-Environment	Functional
G-9	Cornwall Seal Group	Natural Systems	Natural Beauty
G-10	Coast Path Association	Human-Environment	Natural Beauty
G-11	AONB Partnership	Human-Environment	Natural Beauty
G-12	Local Surf School	Human-Environment	Natural Beauty